A Biography of Paul Berg

THE RECOMBINANT DNA
CONTROVERSY REVISITED

A Biography of Paul Berg

THE RECOMBINANT DNA
CONTROVERSY REVISITED

ERROL C. FRIEDBERG
University of Texas Southwestern Medical Center at Dallas, USA

WITH A FOREWORD BY
SYDNEY BRENNER
Nobel Laureate in Physiology or Medicine, 2002

WS World Scientific

NEW JERSEY · LONDON · SINGAPORE · BEIJING · SHANGHAI · HONG KONG · TAIPEI · CHENNAI

Published by

World Scientific Publishing Co. Pte. Ltd.

5 Toh Tuck Link, Singapore 596224

USA office: 27 Warren Street, Suite 401-402, Hackensack, NJ 07601

UK office: 57 Shelton Street, Covent Garden, London WC2H 9HE

Library of Congress Cataloging-in-Publication Data

Friedberg, Errol C., author.

   A biography of Paul Berg : the recombinant DNA controversy revisited / Errol C Friedberg ; foreword by Sydney Brenner.

    p. ; cm.

   Includes bibliographical references.

   ISBN 978-9814569033 (hardcover : alk. paper) -- ISBN 978-9814569040 (pbk. : alk. paper)

   I. Title.

   [DNLM: 1. Berg, Paul, 1926–   2. Genetics--United States--Biography. 3. DNA, Recombinant-- United States. WZ 100]

   QH447

   572.8'6092--dc23

   [B]

                                            2013041976

British Library Cataloguing-in-Publication Data

A catalogue record for this book is available from the British Library.

Printed by FuIsland Offset Printing (S) Pte Ltd Singapore

About the Author

Errol C. Friedberg, M.D., is Emeritus Professor of Pathology at the University of Texas Southwestern Medical Center at Dallas, Texas. He is a Fellow of the Royal College of Pathologists and Honorary Fellow of the Royal Society of South Africa and holds an Honorary Doctorate of Science from the University of the Witwatersrand, Johannesburg, South Africa. He is a recipient of the Rous Whipple Award from the American Society of Investigative Pathology and the Lila Gruber Honor Award from the American Academy of Dermatology. Friedberg is the author of *Cancer Answers, Correcting the Blueprint of Life: An Historical Account of the Discovery of DNA Repair Mechanisms, The Writing Life of James D. Watson, From Rags to Riches: The Phenomenal Rise of the University of Texas Southwestern Medical Center at Dallas, and Sydney Brenner: A Biography.* He is also senior author of the textbook *DNA Repair and Mutagenesis* and has edited and annotated a series of interviews published as *Sydney Brenner: My Life in Science.*

Friedberg, E. C. (1985) *DNA Repair.* W. H. Freeman & Co., New York.

Friedberg, E. C. (1992) Cancer *Answers: Encouraging Answers To 25 Questions You Were Always Afraid To A*sk. W. H. Freeman & Co., New York.

Friedberg, E. C., Walker. G. C. and W. Siede. (1995) *DNA Repair and Mutagenesis.* ASM Press, Washington, D.C.

Friedberg, E. C. (1997) *Correcting the Blueprint of Life: An Historical Account of the Discovery of DNA Repair Mechanisms.* Cold Spring Harbor Laboratory Press. Cold Spring Harbor, New *York, NY.*

Friedberg, E.C. (2005) *The Writing Life of James D. Watson.* Cold Spring Harbor Laboratory Press, Cold Spring Harbor, New York, NY.

Brenner, S., Wolpert, L., Friedberg, E. C. and Lawrence, E. (2005) *My Life in Science*, BioMed Central.

Friedberg, E. C., G. C. Walker, W. Siede, R. D. Wood, R. A. Schultz and T. Ellenberger (2006) *DNA Repair and Mutagenesis* (2nd edition) ASM Press, Washington, DC.

Friedberg, E. C. (2007) *From Rags to Riches. The Phenomenal Rise of The University of Texas Southwestern Medical Center at Dallas.* Carolina Academic Pres. Raleigh Durham, NC.

Friedberg, E. C. (2010) *Sydney Brenner-A Biography.* Cold Spring Harbor Laboratory Press, Cold Spring Harbor, New York, NY.

For my sons
Malcolm, Andrew, Jonathan and Lawrence
&
my grandsons
Mason, Hayden and Levi

History is that certainty produced at the point where the imperfections of memory meet the inadequacies of documentation.

Julian Barnes (author)

Contents

Timeline

1926 (June 30)	Birth date
1940	Berg enters Abraham Lincoln High School, Brooklyn, New York
1943 (November)	Berg Enlists in the US Navy
1944	Berg enters Pennsylvania State (Penn State) University
1946	Berg is discharged from the US Navy and returns to Penn State
1947 (September 13)	Berg marries Mildred (Millie) Levy
1948	Berg graduates from Pennsylvania (Penn) State University
1948	Berg enters graduate school at Case Western Reserve University, Cleveland, Ohio
1952	Berg graduates with a Ph.D. degree from Case Western Reserve University
1952–53	Postdoctoral work with Herman Kalckar at the Institute of Cytophysiology, Copenhagen, Denmark
1953	Postdoctoral work with Arthur Kornberg at Washington University, St. Louis

1955–1959	Asst. Prof./Assoc. Prof., Dept. of Microbiology, Washington University, St. Louis
1956	Berg discovers a new enzymatic mechanism for acetate activation
1958 (September 30)	John Berg born in Palo Alto, California
1957–58	Berg discovers tRNA in the bacterium *E. coli*
1959–60	Assoc. Prof., Dept of Biochemistry, Stanford University
1960–2000	Professor, Dept of Biochemistry, Stanford University
1969–1974	Chairman, Dept. of Biochemistry, Stanford University
1961	Berg attends International Biochemistry Congress, Moscow, USSR
1965	Berg hears a pivotal lecture in Dale Kaiser's graduate course at Stanford
1967–68	Berg takes sabbatical leave at the Salk Institute, La Jolla, California
1968	Berg initiates experiments to generate recombinant DNA molecules
1968 (November, 26)	Berg applies to the *American Cancer Society* (ACS)for a research grant to support this research
1969	Stanford graduate student Peter Lobban independently conceives ideas for generating recombinant DNA
1969	John Morrow joins the Berg laboratory as a graduate student
1969 (May 14)	Berg receives notification of a grant award from the ACS
1969 (March)	David Jackson joins the Berg laboratory as a post-doctoral fellow

1969 (November 6)	Peter Lobban's 3rd examination, Dept. of Biochemistry
1970–1994	Sam, Lulu and Jack Willson Professor of Biochemistry
1970	Berg publishes his first paper on eukaryote molecular biology
1970 (September)	Janet Mertz joins the Berg laboratory as a graduate student
1970	Berg submits a renewal research proposal to the *American Cancer Society*
1970	Robert (Bob) Symons (sabbatical visitor from Australia) joins the Berg laboratory
1970	Morton Mandel and Akiko Higa establish a technique for making *E. coli* more "competent" for transforming foreign DNA.
1971 (June)	Janet Mertz attends a course at the Cold Spring Harbor Laboratory, Cold Spring Harbor, New York
1971 (June)	Bob Pollack telephones Berg about Berg's recombinant DNA work
1971 (Summer)	John Morrow acquires *Eco*R1 endonuclease from Herb Boyer
1971 (September)	Berg writes a progress report on his *American Cancer Society* grant
1971	Douglas Berg (in Dale Kaiser's laboratory) constructs the first plasmid bacterial cloning vector (λ*dvgal*) with Janet Mertz, David Jackson and Douglas Berg
1971–1972	Jackson *et al.* and Lobban and Kaiser concurrently develop the terminal transferase tailing method for joining DNA *in vitro*.
1972 (May)	Peter Lobban submits his Ph.D. thesis to Stanford University and departs for Canada for post-doctoral work

1972 (July)	Lobban writes "thank you" letter to Berg.
1972 (July)	Berg submits paper on generating recombinant DNA to the *PNAS* (Jackson, *et al.*)
1972 (October)	Jackson, Symons and Berg paper published in the *PNAS*
1972 (November)	*Nature* publishes editorial voicing concern about Jackson, *et al.*, paper
1972 (August)	Berg submits a paper (Morrow and Berg) to the *PNAS* on cutting of DNA by *Eco*R1 enzyme
1972 (November)	Morrow and Berg paper published in the *PNAS*
1972 (Fall)	Berg invites Pollack, Hellman and Oxman to organize Asilomar I meeting
1972 (September)	Berg communicates paper on cohesive ends generated by *Eco*RI enzyme to the *PNAS* (Mertz and Davis)
1972 (November)	Mertz and Davis paper published
1972 (November)	NIH Biohazards Committee established
1972	Stanley Cohen begins plasmid experiments at Stanford
1972 (November)	Goodman and Boyer publish sequence of *Eco*R1 cleavage site (communicated to the *PNAS* by Berg)
1972 (November 13–15)	Plasmid meeting in Hawaii, attended by Cohen and Boyer
1972–74	Cohen *et al.* (1972) isolate the drug-selectable bacterial cloning vector, pSC101. They use it to construct, clone and express bacterial intra- (1973) and inter- (1974) species recombinant DNA.
1973 (January 22–24)	Asilomar I conference, Asilomar, California

1973 (May)	Cohen sends paper announcing plasmid pSC101 to Norman Davidson for communication to the *PNAS*
1973 (June 11–15)	*Gordon Research Conference on Nucleic Acids*, New Hampton, New Hampshire
1973 (September 21)	Singer-Söll letter published in *SCIENCE*
1973 (November)	Cohen *et al. PNAS* paper on plasmid pSC101 published
1973–1983	Non-resident fellow, Salk Institute, La Jolla, California
1974 (April 17)	Berg convenes a meeting at MIT to discuss Singer/Söll letter. The "Berg letter" is written.
1974 (May)	Morrow, Cohen, Boyer *et al.* publish a paper in the *PNAS* on cloning Xenopus DNA in plasmid pSC101
1974	John Morrow completes his Ph.D. thesis
1974 (May 24)	Executive Committee of Assembly of Life Sciences names Berg MIT Committee the *NAS Committee on Recombinant DNA Molecules*
1974 (July 18)	Press Conference in Washington, DC announcing the "Berg letter"
1974 (July 19)	"Berg letter" published in *Nature*
1974 (July 26)	"Berg letter" published in *SCIENCE*
1974 (August)	Ashby Working Party established in UK
1974 (September 10)	*NAS Committee on Recombinant DNA Molecules* (MIT group) convenes again and decides on a date for the Asilomar II meeting
1974–1975	Filing of initial Stanford University/ University of California, San Francisco (UCSF) Cohen/Boyer patent applications relating to recombinant DNA.
1975 (January)	Ashby report on recombinant DNA in UK is published

1975 (February 22–23)	Working Panels meet at Asilomar
1975 (February 24–27)	Asilomar II conference
1975 (February 28)	NIH Advisory Committee meets for the first time (in San Francisco)
1975 (April 22)	US Senate Subcommittee on Health hearing on recombinant DNA (Sen. Edward Kennedy, chair)
1975 (May)	Asilomar II report submitted to the NAS for approval
1975 (May 15–16)	NY Academy of Sciences Conference — *Ethical and Scientific Issues Posed by Human Uses of Molecular Genetics*
1976 (June 6)	Asilomar II Report published in *SCIENCE*
1976	Proceedings of NY Academy of Sciences Conference published
1976 (June)	NIH issues initial recombinant DNA guidelines. Cambridge (Mass) City Council Meeting at which Mayor John Velluci proposes a 2-year moratorium
1976	Herb Boyer and Robert Swanson co-found *Genentech*, the first biotechnology company
1976 (Sept. 22)	US Senate Health Subcommittee oversight hearing on recombinant DNA (Sen. Edward Kennedy, chair)
1977 (November 2)	Berg testifies to US Congress
1978 (December)	Revised NIH guidelines released and become effective January 2, 1980
1980	Stanford/UCSF (Cohen/Boyer) patent issued by US Patent Office
1985–2000	Berg appointed Director of the *Arnold and Mabel Beckman Center for Molecular and Genetic Medicine*, Stanford University

1991	Berg chairs the *Human Genome Project Advisory Committee*, Office of Genome Research, NIH
1991	Berg publishes *Genes & Genomes* with Maxine Singer
1992	Berg resigns position at the NIH Office of Genome Research
1992	Berg publishes *Dealing With Genes — The Language of Heredity* with Maxine Singer
1993	Berg joins Board of Directors of *Affymetrix*
1994–2000	Vivian K. and Robert C. Cahill Professor in Cancer Research, Stanford University
1997	Berg publishes *Exploring Genetic Mechanisms* with Maxine Singer
1998–	Berg joins Board of Directors, *Gilead Sciences*
2000	Berg resigns as Director of the *Arnold and Mabel Beckman Center for Molecular and Genetic Medicine* and closes his Stanford research laboratory
2000–	Vivian K. and Robert C. Cahill Professor Emeritus in Cancer Research
2003	Berg publishes *George Beadle — An Uncommon Farmer* with Maxine Singer
2004	Paul and Mildred Berg Hall dedicated in the *Li Ka-Shing Center for Learning and Knowledge*, Stanford University
2004–	Public policy activist for stem cell research

Notable Honors and Awards

To Paul Berg

1959	Eli Lilly Award in Biochemistry
1966	Elected to US National Academy of Sciences
1966	Elected to American Academy of Arts and Sciences V.D. Mattia Prize of the Roche Institute for Molecular Biology
1974	Elected to Institute of Medicine, US National Academy of Sciences
1978	Honorary Doctor of Science, University of Rochester
1978	Honorary Doctor of Science, Yale University
1979	Sarasota Medical Award for Achievement and Excellence
1980	Annual Award of the Gairdner Foundation
1980	Nobel Prize in Chemistry
1980	Albert Lasker Basic Medical Research Award
1981	Elected Foreign Member, French Academy of Sciences
1982	American Association for the Advancement of Science Scientific Freedom and Responsibility Award (1982)
1983	National Medal of Science
1983	Elected to American Philosophical Society
1986	Honorary Doctor of Science, Washington University, St. Louis
1986	National Library of Medicine Medal
1988	American Academy of Achievement
1989	Honorary Doctor of Science, Oregon State University

1991	Honorary Member of the Academy of Natural Sciences of the Russian Federal Republic (1991)
1992	Foreign Member of the Royal Society, London
1995	Honorary Doctor of Science, Pennsylvania State University
1996	Elected Member of Pontifical Academy of Sciences
1999	Max Delbrück Medal
1999	Albany Medical College Theobald Smith Award
2003	Research America Prize for Sustained Leadership
2005	Biotechnology Heritage Award

Acknowledgements

Documenting Paul Berg's life required the assistance of numerous individuals during different phases of the writing. I am especially indebted to Jack Berg, John Berg, Millie Berg, Stanley Cohen, Mel dePamphilis, Jerry Hurwitz, David Jackson, Dale Kaiser, Peter Lobban, Janet Mertz, John Morrow, Suzanne Pfeffer, Philip Pizzo, Maxine Singer, and Charley Yanofsky for interviews, and in particular for the candid and forthcoming manner in which these unfolded. Paul Berg's contributions were beyond the pale. Aside from his unwavering interest and co-operation, he assumed a role that might ordinarily merit co-authorship. Though emphatically requiring his help with the recall of dates and names, I did not anticipate the depths to which Paul immersed himself during my writing — without foisting unwelcome opinions.

Thanks also to Mike Brown, Diego Castrillon, Andrew Friedberg, Rhonda Friedberg, Rene Galindo, Peter Lobban, John Morrow, Nancy Schneider, Eric Olsen, Wei Yang, Maxine Singer and Suzanne Pfeffer for their critical reading of parts or all of drafts in various states of completion. I owe a large debt of gratitude to individuals in the Department of Special Collections at Stanford University for their tireless efforts in retrieving and returning close to thirty linear feet of Berg's archived material. Thanks are also due to the MIT Oral History Program and to the University of California Oral History Collection for invaluable transcripts of interviews conducted under their auspices. Special thanks are due to my wife Rhonda for her moral support during the several years that I was engaged in this effort — not infrequently away from home — and to Sylvia Friedberg and Larry Culp for their generous hospitality.

Last, but by no means least, I thank my many teachers at schools and university in South Africa who first instilled my long held passion for history. In the words of Virginia Woolf, "*I can only note that the past is beautiful because one never realizes an emotion at the time. It expands later, and thus we don't have complete emotions about the present, only about the past.*"

Foreword

A few days after Francois Jacob and I completed the hectic messenger RNA experiments at Caltech in the summer of 1960 I went to San Francisco to see Gunther Stent. He drove me to Stanford University where I was asked to give an impromptu lecture on our experiments. I was taken to a large lecture hall and just before I was due to begin the Biochemistry Department, led by Arthur Kornberg, marched in through a door on my left, and shortly after, the Genetics Department, led by Josh Lederberg, marched in from the right. They all took seats in the front row and I was signaled to begin. My talk was not a success, as few understood it — no enzyme had been purified and no genetic crosses had been done!

After the talk I had a futile argument with Arthur Kornberg about the base composition deduced from the messenger RNA made by the T4 bacteriophage. As the reader will discover, Arthur Kornberg ran the department with iron discipline and all were trained to present their work with precision; my rather slapdash chalk drawings on a blackboard did not fit their style. In later years when Arthur and I served on a committee to found a new university in Okinawa, Japan, I came to appreciate him more, especially for his insistence that all meetings had to end at 5:30 pm so (he and I) could have our evening Martini cocktails!

I had little to do with Paul Berg until much later when he came to talk to me about recombinant DNA and subsequently recruited me to serve on his Organizing Committee for the Asilomar Conference. I had been working on *C. elegans* for about a decade and was dispairing as to the prospect of characterizing the genes. Based on our experience with what we called "steam" genetic engineering, the ability to isolate fragments of *E. coli* DNA during specialized transduction experiments, I decided that I would try to isolate a gene from *C. elegans* by looking for xxvi *A Biography of Paul Berg*

transformants in *Bacillus subtilis*. I chose a gene of carbohydrate metabolism, glyceraldehyde phosphate dehydrogenase, which had been found to have a similar amino acid sequence in yeast, lobsters and humans. I would try to obtain mutants and examine a large number of events to find a lucky break. It was an experiment of desperation and would not have worked at all; because we did not know about introns.

Thus, when Paul Berg told me about the recombinant DNA experiment he was planning, I saw that the technology together with the methods that Fred Sanger was developing at the LMB would create a major revolution in biological research. There was a long journey to what we now have and much is discussed in this biography of Paul Berg, which also relates him and his accomplishments to the remarkable laboratory they created at Stanford.

Professor Sydney Brenner
Nobel Laureate

Preface

By the late 1960's many fundamental aspects of gene function in unicellular organisms such as bacteria and the viruses (bacteriophages) that usurp them as hosts were reasonably well understood. But deciphering gene function in more complex organisms comprising billions of cells organized into multiple functional entities we call organs, was entirely another matter. No one had keys to open such doors, let alone the knowledge to venture through them.

Aware of the enormous boon that bacteriophages had historically provided in understanding how genes from simple organisms work, Paul Berg decided to hang his intellectual hat on achieving a deeper understanding of some of the viruses that infect mice — and you and me. His plan was essentially to join pieces of bacterial DNA to those from a virus called simian virus 40 (SV40) in the test tube and introduce these recombinant (joined) bacterial-viral genomes into mouse cells on the one hand and bacterial cells (such as the common intestinal bacterium *E. coli*) on the other. His primary goal was to determine whether genes contained in these recombinant DNA molecules could function in foreign environments. Specifically, are bacterial genes able to function in higher organisms — and vice versa, and if so, how? Deploying his extensive experience and expertise in the biochemistry of DNA, Berg and his laboratory colleagues devised an ingenious strategy for generating recombinant DNA molecules, work that led to a Nobel prize in 1980 — and that heralded the discipline of genetic engineering. Remarkably, at about the same time others at Stanford University had similar ambitions, though born of different goals, notably the isolation and propagation of individual genes (gene cloning) and their overexpression to yield large quantities of proteins encoded by them. Predictably these events led to some questions about who discovered what and when.

No one had previously introduced recombinant DNA molecules carrying foreign DNA into bacterial cells, and more than a few voiced concerns that such manipulations might lead to unintended and potentially dangerous genetic outcomes. (Recall that *The Andromeda Strain*, by the late Michael Crichton, a techno-thriller novel documenting the efforts of a team of scientists investigating a deadly microorganism, had then recently been published — to wide acclaim.) But Berg and other reputable scientists were excited about the potential benefits to humankind of recombinant DNA technology, including the possibility of gene therapy for devastating hereditary diseases. They believed that the risks of real and imagined biological nightmares were so remote that they did not merit serious consideration. But others argued that it was impossible to categorically state that such experiments were absolutely and unconditionally risk-free. In short order a community of scientists itching to move forward with experiments that fell into the general category of genetic engineering, engaged in impassioned arguments with naysayers who argued vociferously against any and all such efforts. Thus was born the so-called *recombinant DNA controversy*, a controversy of international proportions that generated one of the most momentous debates in the history of science — with Paul Berg at its epicenter.

Over the years the recombinant DNA controversy has spawned an impressive "mini-library" of books, articles in scholarly journals and extensive media coverage (see references at the end of the book.) So why another literary offering now? It goes without saying that no literary rendering of Paul Berg's life can ignore this central episode in the history of biology in which he played seminal roles. Then too, most written accounts of the saga were published many years after the events transpired — some as long as 30–40 years later. Human memory being as fallible as it is, parts of several of these historical accounts are inconsistent and inaccurate with respect to who said what, where and when. Additionally, a book by former NIH Director Donald Frederickson notwithstanding, none of the publications referred to above was authored by scientists, or for that matter anyone else intimately familiar with the discipline of molecular biology.

In view of these shortcomings a significant fraction of the content of this biography is deliberately devoted to direct quotes from documented sources and interviews, a strategy adopted with the specific intention of providing the reader with verbatim accounts of the sort of conflicting opinions and conclusions that so often dog historical records. Thankfully, some of the most penetrating documented interviews by others emerged from forward-looking academic entities, including the Massachusetts Institute of Technology (MIT) Oral History Program, the libraries of the Stanford University Department of Archives, and the Regional Oral History Office of the Bancroft Library at the University of California at Berkeley.

It is important to differentiate between a complete *history* of recombinant DNA and its many consequences, and the recombinant DNA *controversy*, a contentious period between circa 1969–1975, during which those in favor of the technology and those concerned about serious health and other biological risks that may accrue from molecular manipulations of the human genome clashed. This book makes no attempt to cover the entire history of recombinant DNA. Notably, the later period of this era, during which extensive regulatory procedures for the safe use of recombinant DNA technology were promulgated and ultimately initiated, did not engage much of Berg's time and energy. Nor is the massive financial spin-off of gene cloning that followed on the heels of the scientific contributions by Berg and others treated here in any depth.

The reader should be aware too that the recombinant DNA controversy was by no means limited to the United States. However, this book makes no attempt to systematically cover events that transpired elsewhere. In a nutshell, it is an account of an absorbing central controversy surrounding the emergence of this technology, and the players who were central to the hullabaloo from the late 1960's to the mid-1970's.

The enormous promise of the application of recombinant DNA technology to medicine and biology, topics that deserve dedicated texts, is also left to others. Recombinant DNA technology heralded the feasibility of sequencing the entire human genome, comprising billions of nucleotides. The subsequent sequences of the genomes of multiple other organisms, both unicellular and multicellular, now number in the hundreds. These efforts have greatly enriched molecular and evolutionary genetics. Recombinant DNA technology has also facilitated stunning progress in medicine and pharmacology, as well as genetic engineering in animals and plants, and still offers the promise of gene therapy.

Nor is this work intended as a definitive accounting of the life and times of Paul Berg, who, nearing the age of 87 at the time of this writing, remains in good health. As is true of many of the giants of the golden age of molecular biology, ultimate renderings of their lives and critical analyses of their scientific and socio-political contributions must await the passage of time.

In these pages, readers will hopefully encounter the personality and temperament of Paul Berg, an exceptional scientist, public policy spokesperson and writer about various aspects of contemporary biology: a man endowed with the insight, fortitude and passion essential for rescuing 20th century cutting-edge biology from a profoundly threatening impasse that might have seriously harmed the long-standing trust between science and scientists on the one hand, and the public and their governments on the other. Readers will discover too a man who earned the respect and gratitude of a multitude of scientists (and politicians), not to mention the many students and fellows who passed through his Stanford University laboratory.

Errol C. Friedberg
November, 2013

Prologue

From the time that he entered graduate school in 1948 until the late 1960's when he had risen to the position of Professor of Biochemistry at Stanford University, Paul Berg's research was resolutely focused on the study of enzymes and the multitude of biochemical reactions they catalyze. During that extended period he followed in the footsteps of distinguished mentors; Warwick Sakami and Harland Wood at Western Reserve University in Cleveland Ohio; the Danish biochemist Hermann Kalckar at the University of Copenhagen; and Arthur Kornberg at Washington University in St. Louis. Beginning with studies in intermediary metabolism, his research progressed to the biosynthesis of nucleic acids and proteins, an area of biochemistry that spawned the rapidly burgeoning field of molecular biology.

Berg entered the stage of eukaryotic molecular biology resolutely focused on understanding the elaborate mechanisms involved in decoding DNA to yield proteins — the essential workhorses in living cells. But the challenges were daunting. Indeed, when the molecular biology of simple organisms like bacteria and bacteriophage was at full flower and all of biology seemed encouragingly tractable to experimental manipulation, more than a few accomplished stalwarts of the discipline attempted comparable transitions — typically with little to show by way of noteworthy outcomes.

Berg posed two fundamental questions. Does the mechanism(s) of gene expression and transcriptional regulation in higher organisms transpire in essentially the same manner as in simple prokaryotes such as the bacterium *E. coli*, i.e., can bacterial genes be accurately expressed in mammalian cells and retain their functional integrity? And reciprocally, can genes from higher eukaryotes be accurately expressed in bacteria? Parenthetically, he also hoped to establish whether the structure of genes and their arrangement on chromosomes in higher organisms were similar to or radically different from that in prokaryotes.

In the midst of these reflections Berg attended a lecture in a graduate course offered by his Stanford biochemistry faculty colleague Dale Kaiser, a highly regarded microbial geneticist and an authority on bacteriophage lambda (l). Berg was prompted to attend what turned out to be the final lecture of Kaiser's course because its title suggested that Kaiser intended to draw parallels between the biology of bacteriophages and their life cycle in bacteria, and those of viruses and *their* biological interactions with mammalian cells.

Notably and of particular interest to him, Berg learned that like many bacteriophages, some mammalian viruses such as polyoma (Py) virus* and a related entity called simian virus 40 (SV40)** integrate into the genome of host cells. Mammalian cells so affected (mouse cells are a well studied example) frequently manifest features of cancer cells — a phenomenon referred to as *neoplastic transformation*. When grown under laboratory conditions, such cells have offered tantalizing clues to the many complexities associated with cancer cells.

---

*Polyomavirus (Py) (a name that refers to the virus' ability to produce multiple tumors) is the sole genus of viruses within the family *Polyomaviridae*. [http://en.wikipedia.org/wiki/Polyomavirus] The virus (of which there are multiple strains) contains a double-stranded circular DNA genome; that of the commonly used A2 strain is 5292bp in size. Py can persist as a latent infection in a host without causing disease, but may produce tumors in a host of a different species, or one with an ineffective immune system [http://en.wikipedia.org/wiki/Polyomavirus].

**SV40 is an abbreviation for Simian vacuolating virus 40 or *Simian virus 40*, found in both humans and mice. Like other polyomaviruses, SV40 has the potential to cause cancer, but most often persists as a latent infection.

Dale Kaiser. (http://academictree.org/photo/cache.004426.jpg)

The more he pondered initiating a research program in eukaryotic molecular biology, the more Berg realized the imperative of first acquiring the technical contributions to the study of viruses and cancer by spending a sabbatical year in Renato Dulbecco's Salk Institute laboratory in La Jolla, California.

Aside from the anticipation of an exceptional scientific milieu, the geographical proximity of La Jolla to Palo Alto promised a comfortable commute to and from Berg's Stanford laboratory, facilitating frequent contact with his graduate students and post-doctoral fellows. Then too, the greater La Jolla area was celebrated for its scenic splendor and relaxed lifestyle, offering an inviting environment to engage the entire Berg family for a year.

Berg, his wife Millie and his then young son John made the short journey south to La Jolla in the fall of 1967, where they were joined by Berg's long time technical assistant Marianne Dieckmann and by a young French scientist, Francois Cuzin, a recent post-doctoral fellow from Francois Jacob's laboratory at the Pasteur Institute in Paris. In the course of a thoroughly enjoyable and scientifically fruitful year, during which the team learned the ins and outs of handling mammalian cells and some of their viral inhabitants, the Stanford group's preliminary explorations of the interactions of polyoma virus with mouse cells yielded Berg's first research paper devoted to higher organisms, entitled *Induction of virus multiplication in 3T3 cells transformed by a thermosensitive mutant of polyoma virus. Formation of oligomeric polyoma DNA molecules*, published in the *Journal of Molecular Biology* in 1970.[1]

While at the Salk Institute, Berg was particularly excited to learn about entities called *pseudovirions*, virus particles that sometimes co-opt random bits of mammalian host DNA. "I was struck by the phenomenological similarity between pseudovirions carrying bits of mammalian DNA and bacteriophage that had acquired bacterial genes following integration into their genomes. So I began to wonder whether it might be feasible to use pseudovirions to isolate mammalian genes of interest."

On November 26, 1968, Berg submitted a research proposal to the American Cancer Society entitled *Viral Oncogenesis and Other Problems of Regulation*. He wrote:

> For the past fifteen years or so, my research has been concerned primarily with the biochemistry of genetic processes and more specifically with the enzymatic mechanisms of replication, transcription and translation in microbial systems. This approach has uncovered much important information; but, equally important, now we can begin to examine and interpret more complex biologic questions in terms of the concepts developed with the microbial models. How is the expression of genetic information regulated by intrinsic and extrinsic factors in eukaryotic cells? Are the models of repression-induction control of protein synthesis applicable to the genetic organization of higher cells? Or are there new mechanisms and principles to be uncovered?[2]

Berg was well aware of the utility of bacteriophages to transfer genes from one *E. coli* strain to another, and the contribution that this technology had made to our understanding of the molecular biology of bacteriophages, including one called P1. He had in fact used P1 bacteriophage-mediated gene transfer in earlier collaborative work with Charles Yanofsky in the Department of Biology at Stanford University designed to modify the genetic make-up of *E.coli* for other reasons.

Sometime in late 1969 or early 1970, Berg composed an undated, unpublished and undistributed essay[3] — an exercise primarily designed to clarify and organize his thoughts about introducing foreign genes into eukaryotic cells. Excited about the potential for deploying pseudovirions as carriers of foreign DNA, he wondered: "Might pseudovirions serve as efficient vehicles for generalized or specialized transduction of mammalian DNA?" Disappointingly, a quick bit of arithmetic yielded a discouraging answer. DNA fragments transduced by P1 phage can be as large as 2% of the bacterial host. In contrast, polyoma or SV40 pseudovirions can only hold about one ten-thousandth of the mammalian genome. Additionally, the genomes of polyoma and SV40 viruses are about a million times smaller than those of mammals. Consequently, even if every gene in the mammalian genome was amplified as much as ten-fold, the probability that any particular gene would be present in a polyoma pseudovirion was not more than a thousandth that of a bacterial gene in a bacteriophage, and the novel notion of using preudovirions to move genes into and out of mammalian genomes went out the window.

Compounding this logistical challenge Berg appreciated that whereas it was routine practice to screen tens of millions of phage-infected bacteria to cull rare transductants by selection for the product of a phage gene(s) of interest, screening considerably fewer animal cells bereft of these technical advantages would be a Herculean undertaking, even if genetic selection was technically feasible in such cells — which was then not the case. "This 'screening factor'," he wrote, "coupled with the very low gene concentration in virus preparations, would severely reduce the detectability of viral transduction, even without taking other probable variables into account, such as the efficiency of recombination in bacterial compared to animal cells."

Before examining the details of these pioneering experiments and their attendant political ramifications, outcomes that ultimately spawned the so-called recombinant DNA controversy, it is pertinent to step back in time and examine Paul Berg's formative years — as a young boy growing up in New York; as a student at high school where he first became seriously enamored with biology; as a college undergraduate at Pennsylvania (Penn) State University and coincidentally as an ensign in the US Navy during World War II; as a graduate student at Western Reserve University in Cleveland Ohio, where at this early stage of his career he made experimental observations that heralded a rising star in the world of biochemistry; as a postdoctoral fellow in Copenhagen and subsequently at Washington University, St. Louis, where he was one of a trio of investigators that independently discovered transfer RNA (tRNA) in bacteria — and ultimately as a faculty member at that institution and subsequently at Stanford University. These formative events in his professional life are related in the succeeding nine chapters.

Notes

1. Cuzin F, Vogt M, Dieckmann M, Berg P. (1970) Induction of virus multiplication in 3T3 cells transformed by a thermosensitive mutant of polyoma virus. Formation of oligomeric polyoma DNA molecules. *J Mol Biol* **47**: 317.
2. Grant proposal to American Cancer Society dated November 26, 1968. Courtesy of Doogab Yi.
3. Berg P. Unpublished essay circa 1968. Personal communication.

PART I

CHAPTER ONE

Growing up in Brooklyn

A pretty rambunctious kid

New York owns a long history of international immigration. Jewish settlement there dates back to 1654, when a small group of Brazilian Jews fled Portuguese persecution and settled in what was then called New Netherlands. Unsettling times in Russia between the late 19th and early 20th centuries prompted vast numbers of Jews to depart that country, the great majority bound for New York, where by the 1950s they constituted a quarter of the city's population.[1] This massive exodus included Paul Berg's parents, Sarah (nee Brodsky) and Harry Bergsaltz, immigrants from a shtetl in Uman, a district in the Kiev Gubernia, an administrative territorial unit in the Russian Empire.

Many Eastern European Jews threatened by growing anti-Semitic sentiments favored Kishinev, a city to which they escaped and sought work. Indeed, by the year 1900, 43% of the population of Kishinev was Jewish — one of the highest ethnic population densities in Europe. But these numbers afforded little protection from Russian hostility. During the first weeks of April 1903 (around the time of Berg's parents' birth), the city suffered a horrific anti-Semitic pogrom that left scores of Jews dead or severely wounded and houses and businesses plundered and destroyed.[2]

This tense political situation, aggravated by the constant threat of military conscription to the Russian Army, prompted droves of Eastern European Jews to leave their birthplace. The mass departure included Berg's parents, whose odyssey took them to New York to join his mother's

3

half-sister, Rose (they shared the same Russian father), her husband Louis and their four children. Uncle Louis owned a horse-drawn cart and earned his living traveling from neighborhood to neighborhood peddling fresh fruit and vegetables. Notwithstanding their own hardships Rose and her husband provided much valued emotional support — not to mention financial assistance — to Paul's parents during their long and sometimes arduous journey to America.

Harry and Sarah Bergsaltz left their homes the very day after their wedding intent on working their way to Antwerp Belgium, from whence they planned to sail to the United States. Berg's father was just 20 years old — his mother Sarah a year younger. Three grueling years elapsed. Illegally passing through one country after another, the couple was sometimes forced to cross borders covertly. On other occasions they bribed their way to safety. In Kishinev, Harry Berg was able to ply his previously learned skills as a tailor, adding furs to hats, coat collars and full-length coats. Little is known of the details of those harrowing years except that the couple was constantly on the move, working one job after another to sustain themselves. Once in Antwerp, they boarded the vessel *Belgenland* bound for New York.

"It took me a long time to appreciate the enormous courage it must have required to leave their families, survive three years of wandering around Europe and eventually reach America,"[3] Berg related. As if they didn't have enough challenges to contend with, this saga was complicated by the birth of a son, Schmiel, six months before Harry and Sarah reached New York. Imagine if you can the anguish occasioned by the death of their infant not too long after their arrival in the United States — the first of two brothers in the family to die prematurely.

Once settled, Berg's parents shortened their name from Bergsaltz to Berg. Contrary to popular belief, the view that immigration officers at Ellis Island frequently and arbitrarily changed Jewish family names has been debunked in genealogical circles.[4] Officials had no cause to rely on their limited comprehension of the spelling or pronunciation of names uttered by anxious and often confused immigrants right off the ships. They were in possession of formal passenger manifests usually prepared at the port of embarkation in Europe. Furthermore, by the time Ellis Island was fully up and running in 1892, the official apparatus was vast enough to ensure no shortage of translators even for the more obscure European languages. In short, name changes during the immigration process were almost always the result of actions by immigrants themselves.[5] Ellis Island immigration records document the arrival of Aron (Berg's father), Sura (his mother) and baby Schmiel Bergzalz in New York on August 4, 1923. The family's ethnicity is given as Russian/Hebrew and their last place of residence as Orhet, Rumania.

Harry Berg and his wife settled into a multi-family apartment house in Brooklyn, close to their newfound relatives. Harry supported his family plying his skills as a furrier, a lively trade practiced by many New York Jews. Earnings were carefully saved, and in time Berg's father, together with a business partner, acquired sufficient resources to establish their own fur business: *Kaplan and Berg — Furriers*. Paul was born on the last day of June 1926, a day which his mother remembered as "absolutely scorching."

The onset of the Great Depression seriously strained the Berg family's meager resources. "My father was forced to work long hours to keep his business viable," Berg stated. "He left home for work well before the sun rose. Outside of Sundays we hardly saw him before we woke and he typically returned home long after we had eaten dinner. We were not destitute or 'poor' then. But we certainly had to be frugal in our ways. Fortunately my mother had a genius for managing the family affairs and I don't ever recall my brothers or I being hungry or not reasonably well dressed."[6]

Berg has one living sibling, a brother, Jack, a year and half younger than he. Influenced by his older brother's stint in the US Navy during World War II (see Chapter 3), Jack entered the US Merchant Marine Academy shortly after the war ended and graduated with a degree in marine engineering. Following active service in the Navy during the Korean conflict, Jack completed law school at night while holding down a day job in a shipping business. He subsequently cultivated a highly successful career as a specialist in maritime arbitration law, at one time serving as president of the International Maritime Arbitration Organization.

Following along the paths of his older brothers, Berg's second brother, Irving, enlisted in the US Maritime Service as a mid-shipman at the

Maritime Service Academy in Kings Point, New York. While there, he fell ill and was eventually forced to leave the Academy. He was suffering from what was ultimately diagnosed as Crohn's disease, a serious inflammatory condition that affects the large intestine. During his short life, Irving was plagued by multiple episodes of acute intestinal inflammation that sometimes presented as medical emergencies requiring surgical intervention. For the duration of these trying times, Irving was hospitalized at Columbia University. The Berg family did not then enjoy the convenience of owning a car, hence these periods of confinement to hospital required Irving's mother to endure four-hour roundtrips on the subway

Berg (*front right*) with his parents and his two younger brothers, Irving (*front middle*) and Jack (*front left*). (Courtesy of Paul Berg.)

to be at her son's side. She made these trips daily! Even when Irving was in relatively good health, Paul and Jack had little to do with the "baby" of the group. Though never really robust, in the mid to late 1930s Irving enjoyed an extended remission that supported a moderately active life, including gainful employment. But he died in his sleep at the age of 45.

The two elder siblings on the other hand, just a year apart in age, cultivated a close relationship that endures to this day.

"Paul and I were very close growing up — and remain so today," Jack Berg related. "But he was the one who got all the girls! He was very good-looking — and a good athlete to boot and 'went' with two of the most gorgeous girls in the neighborhood — and had several 'steady' girlfriends before he and Millie became an item. I once traveled to Japan on a ship during the war and returned with a kimono for Paul's then current girl-friend, Barbara. But when I presented him with the intended gift he informed me that he and Barbara were no longer an item — and that he had taken up again with Millie! When I was attending law school, a very attractive young woman once approached me and introduced herself as one of Paul's former girlfriends. She asked after Paul and I told her how well he was doing. That was the expectation of everyone who knew him. That Paul would do well!"[7]

A distinctly laid-back personality with a confident and unflappable demeanor, Jack Berg recalled that his older brother enjoyed a lot of atten-tion from their parents — "more so than me," he stated diffidently. "In fact my wife frequently commented on Paul being the favored son in the family. But this sort of talk never really troubled me much."[8] By all accounts Paul's parents were indeed unstinting in their praise of his scho-lastic success and personable demeanor. This attention presumably con-tributed substantially to the healthy measure of self-confidence he acquired at an early age, though his self-assuredness sometimes brought unwelcome consequences. Later passages will reveal some of the (fre-quently amusing) logistical problems that visited Berg as a result of his reluctance to seek advice from others — one presumably born of his fierce sense of independence and self-confidence.

The Berg family resided in a predominantly Jewish neighborhood in the Brownsville section of South Brooklyn where Berg's mother maintained a kosher home — though neither of his parents were seriously practicing Jews. They spoke chiefly Yiddish (and occasionally Russian) at home. This arcane language was the first that young Paul learned to speak fluently. In fact he didn't learn to converse in English until he attended kindergarten. "Before attending kindergarten there was nothing unusual about Jewish kids speaking Yiddish, the language that my family was most comfortable with," Berg related. "But once I became fluent in English I remember being somewhat embarrassed that they spoke that language with a strong Russian/Yiddish accent — and not very well at that."[9]

In due course, Paul's parents moved to a four-family row house in the East New York section of Brooklyn. The family was among a minority in the neighborhood without a private telephone, relying instead on a public phone in a nearby candy store for communication. This situation offered Paul, Jack and other neighborhood kids the relished opportunity to serve as "runners" to summon those being telephoned — collecting modest tips in the bargain.

Berg's childhood was sheltered, free of significant privations and essentially mundanely happy. Street games of all sorts quickly became one

"A boy and his dream"

The Orioles football team. (Courtesy of Paul Berg.)

of his passions. "The neighborhood had lots of kids my age and we played roller skate hockey in the street with sewer covers serving as the goals — and basketball, with bottomless fruit baskets as basketball nets! All in all it was a very carefree time."[10] Berg's love of outdoor games matured into one for sport in general, an enthusiasm he retains to this day. In particular, when he entered his mid-teens, he became a fervent football fan. As a young teenager he played full-bore tackle football with a local team, the Orioles, composed of neighborhood teenagers, and one which had its fair share of wins and losses!

Berg assumed much of the responsibility for keeping the local football scene alive, helping among other tasks to procure football uniforms, pads and helmets from local sporting goods stores. He avidly followed both the college and professional football scenes, his favorite college football team being that from Stanford University, known then as the Stanford Indians.

"I used to follow the progress of the Stanford team with great interest," he related. "The team had a miserable 1939 season, losing every conference game and winning but a single non-conference game." But the legendary college football coach Clark Shaugnnessy arrived in time for the 1940 season. That year the Stanford team was unbeaten and went on to beat Nebraska in the 1940 Rose Bowl.[11] Of course Berg never then dreamed that he would one day spend Saturday afternoons enthusiastically watching the Stanford football team play in their home stadium in Palo Alto, California.

Berg's formal education began at Public School (PS) 219, an elementary school in the East New York section of Brooklyn, an experience that he recollects as being "easy and fun." To be sure, he skipped a grade in deference to his transparent scholastic ability. "I invariably got As for school work, but I was a pretty rambunctious kid and collected Ds for conduct," Berg related. "So my mother was not infrequently called to school to 'visit' the principal."[12] But after he skipped a grade in deference to his rapid academic progress, Berg's conduct improved and he remained at PS 219 until he completed the 5th grade.

During the late 1930s, Berg's family spent a summer in a rented home in Sea Gate, a shore-side gated community close to Coney Island, at the tip of the peninsula that pokes out from the southern part of the borough of Brooklyn. In the early 19th century this attractive area was home to wealthy residents who established stylish summer houses and indulged their love of ocean sailing, in due course constituting the fashionable Sea Gate Association and the elegant Atlantic Yacht Club, a later stopover for sailing aristocrats. Over the years, the area underwent structural shifts that ultimately accommodated the pocket books of middle-class folks — including the Berg family. These later residents included Isaac Bashevis Singer (Nobel Laureate in literature in 1978), musician Woody Guthrie, and opera star Beverly (Bubbles) Sills,* with whom Berg enjoyed a brief boyfriend/girl-next-door relationship — much to his and brother Jack's delight.

Following a further year of elementary schooling at Coney Island's PS 188, Berg entered the nearby Mark Twain Junior High School, an educational institution distinguished for its academic rigor and talented students. Intellectually gifted pupils (an assemblage that included Berg) were afforded the opportunity of completing the 7th and 8th grade curricula in a single year.

"Mark Twain ranks as one of the most exhilarating periods of my life," Berg related. "The kids selected to participate in these rapid advance classes (called 'RA' classes — for rapid advancement) were bright and energetic and I enjoyed them as classmates. There was a general sense of excitement about learning and the curriculum made a huge impression on me."[13]

Berg's scientific bent was both early (developing well before he attended junior high school) and dedicated, especially in biology. Unlike scientific luminaries such as Jim Watson, (whose father nurtured his passionate interest in ornithology) and Sydney Brenner (whose fascination

---

* Beverly Sills rose to musical prominence when as a youngster she became the voice of the "Rinso Song" on the Rinso radio show. "Rinso white! Rinso bright! Happy little washday song" was one of the earliest musical jingles and was used by Lever Brothers to mass market their soap powder. It was rendered by 12-year-old Belle Sullivan, who later took the stage name Beverly Sills and went on to a distinguished career with the New York Opera Company. [http://www.oshawaexpress.ca/story988.html] [http://www.longislandexchange.com/brooklynqueens/seagate.html] [Hargittai- Candid Science II, Conversations With Famous Biomedical Scientists, p. 159].

with living things was first aroused from reading books), Berg's curiosity was primarily piqued by his nurturing educational environment — and by spotting the occasional dead animal in a sufficient state of preservation to take home to dissect! "Biology interested me most; how living things worked," he stated.[14] "At that early stage these interests sometimes translated to an ambition to become a doctor. Though I didn't think of becoming a practicing physician as much as doing medical research."[15]

Berg's parents were not "bookish." "I can't remember ever seeing either of them reading books," he stated. "They read newspapers, typically in Yiddish. Books were a novelty in my family. But stimulated as I was at school, I read all the time." As a junior high school student Berg devoured Paul de Kruif's classic tale *Microbe Hunters*, an engaging non-fiction novel that relates stories of some of the heroes of microbiology, as well as Sinclair Lewis' popular book *Arrowsmith*. (It's remarkable how many other exceptional life scientists born in the early 20th century can trace a virtually identical progression of their reading interests in biology. Watson and Brenner are good, but by no means exclusive, examples).

Well pleased with Paul's scholastic aptitude and achievements, Berg's parents left well alone, never pointing him in any particular academic direction. "They never queried me about what I wanted to be. They took simple pleasure in my scholastic success, an attitude true of many immigrant Jewish families. Education was highly valued."[16]

Berg attended Abraham Lincoln High, a public school in the Brighton/Sheepshead Bay section of Brooklyn. There he experienced a most decisive educational experience that built indelibly on the formative years of junior high school. Built in 1929, Lincoln High has over the years graduated a singularly impressive cadre of celebrated professionals, including three Nobel Laureates — Arthur Kornberg (Nobel Prize in Physiology or Medicine in 1959 and Berg's future scientific colleague and mentor), Berg himself (Nobel Prize in Chemistry in 1980) and Jerome Karle (Noble Prize in Chemistry in 1985). The school also boasts an imposing list of other notable personalities, including musician Neil Diamond, actor Mel Brooks and authors Arthur Miller and Joseph Heller — not to mention a slew of celebrated athletes.[17]

Berg (and many other students who entered the halls of Abraham Lincoln High) particularly revered a woman staff member named Sophie Wolfe. "Sophie was one of the more important figures in my life in terms of motivation; certainly while I was at high school," he stated.[18] Wolfe was not a formally trained teacher. Her official responsibility was to supervise the school supply room that stocked various materials (a variety of models, chemical equipment and microscopes) regularly used during science lectures. She also orchestrated and managed the so-called *Biology Club*, an informal group of student stalwarts that met most days after classes to pursue their interests in science, which Berg enthusiastically joined on many afternoons. Notwithstanding her lack of formal training, Wolfe was an adept practitioner of the Socratic mode of teaching. When asked a question she would often respond with one of her own, designed to steer the enquiring student in a constructive direction when searching for answers. And if she thought that a question could be usefully addressed by conducting a modest laboratory experiment, she would go so far as to suggest one. "During the two or three years that I was in involved in the Biology Club Sophie constantly stimulated us," Berg stated.[19]

Abraham Lincoln High ultimately named one of the main school building wings after Sophie Wolfe. The school also designated three floors of the wing after its three Nobel alumni, all of whom had come under Wolfe's exceptional influence at one time or another. Berg retains fond and admiring memories of his days at Lincoln High — and especially of Wolfe, to whom he is indebted for his strong foundations in science.

Flooded with vacationing families and visitors from throughout the city, Sea Gate became a veritable paradise for teenagers during the summer months. "There was certainly no shortage of girls," Berg recalled with twinkling eyes, "and no shortage of entertaining ways to indulge one's teenage fantasies." A "gang" of about a dozen (or more) youngsters of both sexes spent virtually the entire summer together, organizing and attending parties (on a regular basis), playing ball games together, and so on. The group practically lived on the beach, which ran the entire length of Sea Gate, often walking or riding their bicycles up and down the expansive boardwalk. One could occasionally view big cruise ships passing on their way to dock at New York harbor. Berg recalls seeing the

Sophie Wolfe (*seated*) with Berg (*right*), Jerome Karle, Nobel Laureate in Chemistry, 1985 (*center*), and an unknown individual, perhaps the school principal (*left*). (Courtesy of Paul Berg.)

fabulous *Queen Mary* and the great French cruise liner *Normandie* docked there.[20]

Coney Island,† (a misnomer since the area is really a small peninsula that hangs from the southernmost edge of Brooklyn) the most famous beachfront locale in the area, was a frequent "hangout" for children of all ages.[21] Often called the "playground of the world," Coney Island has served many roles in the lives and imagination of New Yorkers and visitors. The area enjoyed brief economic stability in the late 1890s and early 1900s, the heyday of Coney Island's famed amusement parks. The rides and concessions were then staples of the local economy. But good times came to an end with the 1964 closing of Steeplechase Park, the last of the major amusement parks in the area.

---

† Like the other Long Island barrier islands, Coney Island once teemed with rabbits and the word "coney" (or "cony") was in fact a popular alternative for "rabbit" at one time. [http://www.vanalen.org/competitions/ConeyIsland/background.htm] [http://en.wikipedia.org/wiki/Coney–Island]

Nathan's — a favorite hangout on Coney Island. (www.nathansfamous.com)

Nathan's Famous‡ original hot dog stand was another Coney Island favorite. Opened in 1916, the place quickly became an area landmark.[22] Berg and his pals frequented the premises on a regular basis, often on bike rides home from school and especially during the annual July 4 Independence Day celebrations, which have featured an annual hot dog eating contest since the facility opened almost a century ago. In recent years this hot dog consumption gala ("feeding frenzy" might be a more apt descriptor) has attracted national television coverage. The 2011 winner devoured 62 hot dogs and buns in the space of 10 minutes before an audience of about 30,000 people!

But life as a teenager was not all school and play. Berg was expected to help out at his father's fur business on an occasional basis, sometimes after school, but more typically during weekends and school vacations. "Of course I was given the most menial sort of work one could imagine," he

‡ Nathan's began as a nickel hot dog stand in Coney Island in 1916 and bears the name of co-founder Nathan Handwerker, who started the business with his wife Ida. He was encouraged by the singing waiters and subsequent Hollywood stars Eddie Cantor and Jimmy Durante to go into business in competition with his former employer. In so doing, Handwerker undercut the man by charging 5 cents for a hot dog instead of 10! [http://en.wikipedia.org/wiki/Nathan's_Famous]

related. "When animal skins arrived at the shop to be processed for their fur they were first stretched by nailing them to a huge board. One of my jobs was to remove the nails when the stretching process was completed. I hated the entire ordeal! I hated the smell — and I particularly hated the fur that constantly irritated my eyes and nose."[23] When he was older (and presumably more reliable from his father's point of view) Berg loaded the finished fur garments onto a pushcart and delivered them to nearby locations.

Eager to earn money independently, during the summer of 1942, Berg secured a job as an office boy in a New York-based company called National Screen Service (NSS), where senior executive Jack Levy handled the distribution of promotional materials for the film industry. As that summer approached, Levy's 15-year-old daughter, Mildred (Millie), joined the ranks of New York teenagers aspiring to earn a little cash while school was out. Millie's doting father acquiesced in principle. But determined to keep a watchful eye on his attractive young daughter, he resolutely informed her that NSS was the only place in the city at which he would permit her to work. So Millie spent her summer delivering mail to offices in the 14-story NSS building. Never shy in the pursuit of female companionship, young Berg did not fail to notice the new "mailman" and promptly introduced himself. In due course he invited the young lady to lunch!

Theirs was not a love-at-first-sight relationship. Nor did Paul and Millie see one another all that frequently during or after that fateful summer, if for no other reason than the handicap of geographical separation. "Millie lived in Howard Beach, a section in Queens where JFK airport is today — and the very southern end of Brooklyn is pretty much as far as one can get from Queens," Berg related.[24] While only occasionally in touch during the remainder of that year, the pair resumed their relationship the following summer when Berg worked for the War Production Board, and his spirits soared when Millie invited him to be her high school prom escort. What romance there was simmered after Berg left for college and during his subsequent tour in the Navy.

Notes

1. http://en.wikipedia.org/wiki/Demographics_of_New_York_City#Jewish.
2. http://en.wikipedia.org/wiki/Chişinău#History.

3. Istvan Hargittai, Candid Science II, Conversations With Famous Biomedical Scientists, p. 1.
4. http://tracingthetribe.blogspot.com/2008/03/ellis-island-myth-and-fact.html.
5. *Ibid.*
6. ECF interview with Paul Berg, April 2011.
7. ECF interview with Jack Berg, Sept. 2012.
8. *ibid.*
9. ECF interview with Paul Berg, January 2012.
10. *ibid.*
11. ECF interview with Paul Berg, April 2011.
12. *ibid.*
13. *ibid.*
14. *ibid.*
15. *ibid.*
16. *ibid.*
17. http://schools.nyc.gov/ChoicesEnrollment/High/Directory/school/?sid=1494.
18. ECF interview with Paul Berg, April 2011.
19. *Ibid.*
20. *Ibid.*
21. http://www.vanalen.org/competitions/ConeyIsland/background.
22. http://en.wikipedia.org/wiki/Nathan%27s_Famous.
23. ECF interview with Paul Berg, April 2010.
24. *ibid.*

CHAPTER TWO

The Essential Paul Berg

A glimpse of the contemporary man

One of the many articles on Berg published in his later years states: "Photographs at Stanford show an athletic man with short wavy hair and a broad distinctive smile [sometimes with just half of his mouth], given to wearing aviator eyeglasses and shirts with rolled up sleeves and button-down collars."[1] Aged 86 at the time of this writing, Berg still commands a striking physical appearance, one that emphatically belies his years. Other than the use of a cane for walking, a relatively recent burden of repeated surgeries on his right leg for a tumor that has defied precise diagnosis since its appearance in 2006 (see Chapter 30), his frame is erect and trim and he sports a (nearly) full head of dark hair — marginally peppered with gray. Some of this good fortune may reflect a lucky draw in the gene pool lottery. The rest he likely owes to his life-long devotion to physical activity, particularly tennis in his later years, a game at which he modestly states he "tried to excel." Berg's tennis partners and opponents were usually friends and scientific colleagues at Stanford, such as biologists Charley Yanofsky and Robert (Bob) Schimke and physicist Steven Chu (who recently served as Secretary of Energy in the US government), and while he would likely not identify himself as overtly competitive, it is said that Berg hated losing!

Berg offers an unassuming and unpretentious demeanor, traits not always encountered among superstars in the world of academia. Most celebrities learn to acquire a mantle of modesty when the occasion requires, but Berg communicates unaffected modesty when shrugging off the many honors and awards (including the Nobel Prize in Chemistry, the National Medal of Science and a clutch of honorary doctoral degrees from prestigious universities) that have come his way during more than half a century of biomedical research. No certificates grace the walls of his office or home. "That's not what you work for," he once told an interviewer.[2] "Satisfaction comes in the ensemble work. That's the heady party of it. The audience's applause is just an added attraction."

An Emeritus Professor since 2000, Berg occupies a small office in the bowels of the Beckman Center for Molecular and Genetic Medicine, a research enterprise of which he was a guiding force for its creation — and the founding director (see later). There he is still frequently sought to lend his vast experience in academic and public policy affairs to the Stanford Department of Biochemistry, the medical school — and sometimes the greater university. Though officially retired from the demanding life of an active scientist directing a busy research laboratory, Berg continues to monitor the progress of contemporary biology by reading the scientific literature and regularly attending seminars at the medical school and elsewhere. Still, he finds time to cultivate both his eclectic taste in non-scientific literature — an appetite too long neglected in his view — and a long-standing and passionate interest in collecting art by contemporary American painters. He also continues to serve as a consultant to several thriving private biotechnology companies in the San Francisco Bay area. "At this stage of my life, there is no shortage of things that grab my attention and pique my interest,"[3] he stated.

Berg's Stanford colleagues, in particular the former dean of the medical school Philip Pizzo, rave about his continued involvement with the medical school and the greater university. "Like everyone in medicine and science I certainly knew Paul Berg's name well before I came to Stanford," Pizzo commented. "But as I began to speak with and get to know him, the most amazing thing to me — an impression that has become more and more refined since — is the incredible depth of this person. Here is somebody who is sort of in the ethereal clouds of greatness to me. Yet he doesn't mind sitting down and discussing issues out of pure interest. In my view Paul is responsible for much of what's taken place at Stanford Medical School since 1959."[4]

Pizzo is especially complimentary about Berg's efforts to bridge the long-standing intellectual chasm between the clinical and basic sciences at Stanford Medical School. "Regardless of the fact that he is someone who has lived strictly in the world of basic science, he sees the value of clinical medicine as something central to the medical school's mission and has energetically embraced fundamental ties between the clinical and basic sciences," Pizzo commented. "He is also more than just a participant in discussions about medical school affairs. Indeed, in that capacity I have seen Paul be highly critical of people or ideas that are at the edge of — let's say, intellectual merit. At 86 he still has the ability to distinguish between excellent things — and not so excellent things; excellent people — from not so excellent people."[5]

Such sentiments are widespread on the Stanford campus. Lucy Shapiro, a highly regarded and internationally recognized biologist, arrived in Palo Alto in 1989 as head of the Department of Developmental Biology, a new basic science department in the Beckman Center (see Chapter 28). She first met Berg during her own graduate student days in Jerard (Jerry) Hurwitz's laboratory at Albert Einstein University. Aware of Pizzo's unfettered admiration of Berg, Shapiro, a good friend and scientific collaborator for many years, forthrightly offers a more measured appraisal. "He's a complicated man in my view," she stated. "If he doesn't like someone he lets him/her know — and he has strong principles and strong beliefs that he will vigorously defend. To deal effectively with Paul and not be snowed under by him one has to have a certain measure of self-confidence."[6] Maxine Singer, another long-standing scientific colleague, friend and co-author of several books with Berg also acknowledges some of the more complex features of Berg's personality, including the difficulty in "getting a word in edgeways when conversing with him."[7]

"Paul is very smart," Shapiro continued. "But he also has that extra dimension of vision, not only in science, but also for himself. He does what *he* wants to do and he goes where *he* wants to go. He would tell Arthur Kornberg what *he* wanted to work on, not what Arthur thought he should work on. He can be stubborn as a mule and very dogmatic. If he wants to do something, nothing can stop him."[8]

Berg thrives on discourse, especially about science, though he brings as much passion to conversations about books, art, politics, ethics — and sport. When in this animated state he enthusiastically dominates conversations, calmly brushing aside unwanted interruptions. His eyes sparkle and he gesticulates freely — smiling all the while — typically with just one side of his mouth. When recalling unpleasant or distasteful events his eyes tend to lose their warmth and his face is set. As later pages will reveal, those who have encountered Berg's ire likely remember it well. But he bears few if any grudges.

Paul Berg's life has been visited by both personal triumphs and tragedies. As recounted later in this book (see Chapter 30), his son John was diagnosed with a B-cell lymphoma in his early middle age, and Berg himself has fought a long and difficult battle with a mysterious tumor on his right leg that has defied precise diagnosis (see Chapter 30). Fortunately the tumor does not appear to be life threatening at the time of this writing. But the triumphs have been both bountiful and spectacular: a celebrated career at one of the world's most prestigious universities during a golden age of science — ultimately acknowledged with a Nobel prize; an opportunity to make a historic contribution to and subsequently formalize one of the most important scientific advances in the history of biology — and in his later years unanticipated personal wealth, affording him and his wife, Millie, the pleasure and fulfillment of modest philanthropy and of making their mutual passion for art a reality.

With this snapshot of the man in mind, the table is set to examine the years of scientific challenge and the many triumphs that visited Paul Berg's life.

Notes

1. Lightman A. (2006) *The Discoveries: Great Breakthroughs in 20th-Century Science, Including the Original Papers.* Vintage.
2. PB interview by Istvan Hargittai, Candid Science II, 2002.
3. ECF interview with Berg, April 2011.
4. ECF interview with Philip Pizzo, August 2012.
5. *Ibid.*
6. ECF interview with Lucy Shapiro, August 2012.
7. ECF interview with Maxine Singer, February 2012.
8. ECF interview with Lucy Shapiro, August 2012.

CHAPTER THREE

College — and World War II

Proudly in uniform

Berg was 15-years old when the surprise Japanese attack on Pearl Harbor shocked the world. He well remembers that day. "The school was assembled in a large auditorium to listen to President Roosevelt on the radio condemning this 'act of infamy' — and then declaring war on Japan," he related.[1] Berg followed the war closely, and before graduating from high school he firmly decided to join the Navy as soon as he turned 17. From the very outset he harbored heroic aspirations to become a fighter pilot in the Navy Air Corp, a choice perfectly suited to his personality and temperament!

Paul's father made a modest living in his fur business — enough to keep food on the table and his family adequately cared for, but insufficient to encourage enrollment at any college that his eldest son's heart might desire, regardless of his obvious academic promise. So Paul stayed close to home, enrolling in the Bronx campus of the City College of New York (then one of the most distinguished universities in the country) where tuition was free if one could muster the required high school grades. Notwithstanding his protracted interest in science, especially biology, as the college years approached Berg's thoughts turned to chemical engineering as a career choice. "I don't remember exactly what motivated my interest in chemical engineering," he reflected. "At the time I thought of it as something practical; something that one could count on to make a reliable living."[2]

An unanticipated logistical nightmare brought an abrupt halt to the City College of New York experience after a mere three days, when Berg (belatedly) realized that commuting back and forth between home at the southern tip of Brooklyn and City College in upper Manhattan consumed about five hours! As we shall see in later chapters, this is but one (likely not even the earliest) example of Paul Berg's bent — presumably born of the sublime self-confidence and fiercely independent streak alluded to earlier — of charging ahead with plans and decisions, sometimes momentous ones, without seeking much (if any) advice from others. He subsequently enrolled in the Brooklyn campus of City College of New York, situated much closer to home. By then he had educated himself about what chemical engineering really entailed — and fatefully decided that this vocation was not for him. He elected a biology major instead.

Coming on the heels of his stimulating experience at Abraham Lincoln High School, biology as taught at Brooklyn College was disappointing. "It was the standard classical biology education of the day, in which one dissected some little pickled animal and made elaborate drawings of everything," he said dismissively. "Totally, totally boring! So I began to turn my thoughts to another career goal that I might enjoy — at least until the time when I expected to be called to military service."[3]

One of Berg's Sea Gate cronies was Jerry Daniels, a close friend who attended Pennsylvania (Penn) State University. At Daniels' recommendation Berg acquired the Penn State handbook and catalogue. While assiduously perusing the biology courses, he was intrigued to discover that the school included a Department of Biochemistry. So there and then he made a decision to study biochemistry at Penn State, blissfully unaware that long before the discipline of biochemistry acquired the lofty academic status it presently enjoys, the subject was frequently practiced and taught in agriculture schools (as was the case at Penn State), where it was heavily focused on analytical chemistry and on preparing most of its graduates for careers in the pharmaceutical and food industries. "Real" biochemistry was largely taught and practiced in medical schools, where it typically bore the moniker "physiological chemistry." Berg spent the summer of 1943 working at the War Production Board (WPB)* in New York before enrolling at Penn State. Eugene (Gene) Rifkin, another of Berg's close friends while growing up in Sea Gate, had also decided to attend Penn State and the pair roomed together. Both were determined to join the Navy as soon as they turned 17.

Berg formally enlisted in the US Navy on November 7, 1943, four months and a few days after his 17th birthday. Rifkin followed suit a few months later. Berg might have made this patriotic gesture earlier but for concern about his long-standing habit of biting his fingernails to the quick. Whether this was factually accurate or not, he had somehow gained the impression that the Navy examined the finger nails of enlistees in the flight program, interpreting unusually short nails as a sign of nervousness — clearly not the stuff of fearless pilots! To avoid such an ignominious outcome Berg regularly painted his nails with iodine, the unpleasant taste of which prompted the self-discipline to grow his nails to a less revealing state. "When asked to stretch out my arms at my induction physical I silently prayed that nothing would be said about my fingernails'," he related.

Berg had expressed a preference for assignment to the Naval Flight Program (the V5 program†) and was assigned for pre-flight training at Middlebury College, Vermont. But after learning that it was standard Navy policy to assign sailors to naval training at colleges and universities at which they were already enrolled, he arranged to be transferred to the V-5 program at Penn State. Now proudly in uniform, Berg and Rifkin roomed together once more. The pair began military training in

---

*The War Production Board (WPB) was established as a government agency on January 16, 1942, by executive order of President Franklin Roosevelt. The purpose of the board was to regulate the production of materials and fuel in the United States.
†On April 15, 1935 Congress passed the *Aviation Cadet Act*. This set up the Volunteer Naval Reserve class V-5 program to send civilian and enlisted candidates to train as aviation cadets. Candidates had to be between the ages of 19 and 25, have an Associate's degree or at least two years of college, and had to complete a Bachelor's degree within six years after graduation to keep their commission. Training was for 18 months and candidates had to agree to not marry during training and to serve for at least three more years of active duty service. [http://en.wikipedia.org/wiki/Aviation_Cadet_Training_Program_(USN)].

In uniform. (Courtesy of Paul Berg.)

March 1944. But Paul's fantasy about becoming a naval fighter pilot ace and knocking Japanese planes out of the skies was not to be realized. The Navy cut back its pilot training program and after completing V-5 training at Penn State, both Berg and Rifkin graduated to midshipman school at Fort Schuyler, on Long Island, New York, for officer training in the Naval V-12[‡] program.

Named in honor of Major General Philip Schuyler of the Continental Army, Fort Schuyler is a preserved 19th century fortification in the Bronx, which among other things housed the New York Maritime College.[4] Berg and Rifkin were what the Navy then obliquely referred to as "90-day wonders" — two of thousands who attended midshipman

---

[‡]The V-12 Navy College Training Program was designed to supplement the force of commissioned officers in the United States Navy during World War II. Between July 1, 1943, and June 30, 1946, more than 125,000 men were enrolled in the V-12 program in 131 colleges and universities in the United States. [http://en.wikipedia.org/wiki/V-12_Navy_College_Training_Program] [http://www.splinterfleet.org].

school for about three months and were then promptly commissioned as ensigns (formally equivalent to the rank of second lieutenant in the other armed services). The pair parted company when as newly minted officers Rifkin was assigned to duty in the Pacific while Berg was dispatched to the deck officer training facility at Key West, Florida. They would resume their friendship at the war's end when both returned to Penn State to complete their college education.

At Key West, officers were trained for duty on a class of ships called submarine chasers (subchasers), the primary function of which was to protect larger naval vessels or merchant ships from attack by enemy submarines. Subchasers were built and launched in great haste at the outset of World War II to hunt and deter German U-boats, which by 1941 roamed the Atlantic essentially unopposed, sinking British and American merchant vessels with complete abandon and with no regard for neutrality.[5] In the summer of 1942, German U-boats sank more ships and took more lives than were lost at Pearl Harbor, and the U-boat menace became a top priority for the Navy, even at the cost of delaying the American response to Japanese aggression in the Pacific.

Over 40,000 men served on subchasers in World War II. The patrol coastal (PC) ships to which Berg was assigned were 173-foot steel hulled, diesel-powered craft designed and first used in World War I. The ships were fitted with one fixed 3-inch cannon and a variety of depth charges for encounters with submarines. They also carried machine guns and anti-aircraft guns. By the end of WWII, the PC owned the proud record of having sunk close to 70 German U-boats.

"I don't think many people appreciated how much American shipping was lost to the German submarines (U-boats) patrolling the Atlantic and wreaking havoc in the Gulf of Mexico," Berg related. "Britain came very close to losing the war because supplies from the United States were so seriously compromised by German U-boats. The subchasers' job was to reduce that devastation." In addition to patrolling the seas for enemy submarines, subchasers led landing craft onto assault beaches, protected them from enemy fire, fought off air attacks, swept for mines and laid down smokescreens. The ships carried crews of about 40–50 sailors and 3–4 officers.[6]

A subchaser crew. Berg is second from the *left* in the row of five officers (*middle*). (Courtesy of Paul Berg.)

Berg remained on active naval duty for about 14 months before the nuclear attacks on Nagasaki and Hiroshima prompted the Japanese surrender. His first subsequent assignment was to help sail subchaser PS 1119 (which had operated in the Pacific) from San Pedro California to Charleston South Carolina for decommissioning. Before proceeding with this assignment he took time to meet up with his brother Jack, then a midshipman in training at the San Mateo California Maritime Academy. The pair hitchhiked down the California coast, stopping every now and then to take in the spectacular vistas. Prompted by a shared appreciation of the mighty Stanford football team, one stopover afforded the two brothers a visit to the Stanford University campus, where they admiringly viewed its many open spaces and the splendid formal gardens and manicured lawns fronting Palm Drive, the main university entrance.

For these avid football fans, visiting Stanford stadium (which in 1927 boasted a seating capacity of 85,000 — 20,000 seats more than it presently holds) was nothing less than awe-inspiring! At that time Berg was in fact harboring the hope of transferring from Penn State to Stanford when discharged from the Navy. But this goal was thwarted by the logistical complexities associated with transferring credits from one institution to the other. Berg had to wait about 15 years before Stanford University became a mainstay of his existence.

Berg made the sea voyage from California to South Carolina by sailing down the Mexican coast and traversing the Panama Canal and the Caribbean Sea to the ship's assigned anchorage in Charleston — where it was to be mothballed. As a junior officer on the ship, Berg took turns at deck watch and temporary command of the vessel for six-hour shifts. One of his ignoble episodes on that assignment transpired late one night when, as officer on the deck, he spotted a bubble track with a perfect lead angle moving directly towards the broadside of the ship. "Believing it could be a torpedo track, I immediately sounded a general alarm and according to required protocol turned the ship 90 degrees to present a smaller target to the oncoming stream of bubbles," he related. "When the awakened captain arrived on deck he quickly pointed out to me that the bubbles came

'Eagle-eye' Berg (1944 - 45)
searching for Nazi subs in the Caribbean

(Courtesy of Paul Berg.)

from nothing more than a school of dolphins — and promptly returned to his quarters."[7] But having duly followed standard operating procedure, Berg was spared any reprimand for waking his captain!

Seasickness was a frequent problem on the small subchasers, especially in foul weather or when the bridge was enveloped in clouds of engine smoke — not an infrequent event. Berg recalls an earlier voyage during which the seas were rough enough to sicken the entire crew, including experienced sea-worthy veterans. "During such adventures one kept a bucket close at hand," he wryly noted.[8]

Most if not all of Berg's teenage friends enlisted in the war effort. At least one, Stanley Greenberg, whom Berg refers to as "a very close friend," was killed in action shortly before the war ended. Berg had been primed and trained for the ultimate invasion of Japan, an initiative that was rendered moot by the atomic bombing of Hiroshima and Nagasaki. Typical of young men eager to experience combat — a mentality as old as war itself, Berg admits to a palpable sense of "disappointment" that

Millie Berg while Berg was in the Navy. The inscription reads: "To my favorite sailor, Mil." (Courtesy of Paul Berg.)

World War II ended before he had any direct experience with it. "It took a while for that sense of 'loss' to disappear," he commented. With the war over, Berg and Rifkin returned to Penn State. They occupied the same temporary barracks, but after a year the duo eagerly accepted invitations to join the Beta Sigma Rho fraternity, not so much wishing to be frat boys as to enjoy the pleasure of "escaping the 'boonies' and the mud-drenched temporary barracks that had been erected to house returning veterans."[9]

Another opportunity for experiencing the "honor and glory" of the battlefield arose when the Korean War broke out in 1950 and Berg, then a graduate student at Western Reserve University (see next chapter), was called to active duty. Now most reluctant to interrupt his graduate education (even for a war), Berg informed biochemistry department chairman Harland Wood of this unexpected development and, following an interventional letter by the School's Dean to the military, Berg's deployment was deferred.

****

Blessed with a becoming physical appearance and a self-assured personality, Berg experienced little difficulty in endearing himself to members of the opposite sex. But Millie Levy remained very much in his consciousness. While he was in the navy, Millie was training at a nursing school in New York. The couple corresponded back and forth, but essentially both were heavily preoccupied with work. "On several occasions my father told me that he had received notes from Paul enquiring about my whereabouts and requesting that he deliver letters to me," Millie Berg recounted. "I wrote back occasionally, and when he was home for the infrequent weekend we went out together. But it was all very casual. Sometimes I visited him to attend Penn State student functions. By then we had both matured considerably and eventually we realized that our long distance relationship was both logistically complicated and tedious — so we decided to marry."[10]

Millie's mother, whom Millie describes as having been "unyieldingly traditional," was not at all pleased about her daughter's intention to marry Paul Berg, who seemed not to represent the sort of man she hoped her daughter would spend the rest of her life with. Besides, she was of the firm opinion that a wife's place was in her home looking after her husband and raising a family, not working (a fairly typical attitude in the US in the late 1940s). She was also decidedly displeased about Millie's intention of working as a nurse, an undistinguished career far beneath her taste. In fact, her discontent was such that she petulantly let it be known that she might absent herself from her daughter's wedding! "When we set the date for the wedding, we resolutely informed her that the event would proceed — whether she appeared or not," Millie firmly related.[11] It did proceed, on September 13, 1947. There was no grand celebration. The couple married quietly in the house of a local rabbi with just their immediate families in attendance — including Millie's mother!

Paul and Millie engaged to be married. (Courtesy of Paul Berg.)

Following a brief honeymoon, the newly-weds moved to State College, the town that accommodated Penn State University — and one described by Berg as "a dinky little place of about 3,000 people!" The couple resided in a rented room a comfortable walking distance from the campus and "downtown."[12] A car was well beyond their financial means and the nearest hospital was a good 10 miles away, so Millie abandoned thoughts of fulltime nursing and busied herself as a Red Cross volunteer. She also lent her nursing skills to assisting several homebound invalids and to helping a resident young mother take care of a newborn infant. Predictably, none of these experiences were especially gratifying to a trained nurse eager to embark on a professional career — nor did they pay appreciably. Hence, within a short time Millie returned to New York, where she worked at Coney Island Hospital, an easy walk from the home of her parents-in-law.

Millie was especially fond of Paul's parents, particularly his mother, and happily resided with them during this period of temporary separation from her husband. "Paul's mother was a wonderful mother-in-law who looked after me very lovingly," she related. "She was the prototypical Jewish mother — but I thoroughly enjoyed the attention. She was an extremely generous, angelic person — very giving and well loved by many."[13]

Berg made the tedious 7–8 hour train journey between State College and New York almost every weekend, gratefully accepting the occasional "lift" from anyone who happened to be driving up to that city. He was pleased with his last two years at college. Regardless of the fact that the Department of Biochemistry at Penn State dwelled in the School of Agriculture, he relished his first serious introduction to biochemistry, in particular finding his way around a research laboratory, and he became increasingly intent on becoming a professional biochemist.

At the end of his junior year, Berg held a paid summer job at the General Foods Corporation in Hoboken, New Jersey, performing routine analytical work in their food technology research laboratories. The following year he acquired a similar position, again working largely in the area of food technology, this time at the Lipton Tea Company, also in Hoboken. Both experiences helped pay expenses that arose at Penn State, as well as travel expenses for weekend trips to New York to be with Millie.

During these summer experiences, Berg soon recognized that employees with bachelor degrees were generally told what to do by others, whereas those giving them direction typically held doctoral degrees! He was sufficiently impressed by this hierarchical observation that he determined that if he was able to carve out a career in biochemical research he was going to be among those calling the shots in the laboratory — not those taking orders from others! "By the end of my junior year in college I had unequivocally decided to enter graduate school in biochemistry and acquire a PhD," he stated.[14] The single remaining question was where, one that turned out to be more complicated than he had hoped!

Notes

1. ECF interview with Paul Berg, April 2011.
2. *Ibid.*
3. *Ibid.*
4. http://en.wikipedia.org/wiki/Fort_Schuyler
5. http://www.splinterfleet.org
6. ECF interview with Paul Berg, April 2011.
7. *Ibid.*
8. *Ibid.*
9. *Ibid.*
10. ECF interview with Millie Berg, April 2011.
11. *Ibid.*
12. ECF interview with Paul Berg, April 2011.
13. ECF interview with Millie Berg, April 2011.
14. ECF interview with Paul Berg, April 2011.

CHAPTER FOUR

Western Reserve University

Enzymology, intermediary metabolism and the trials of learning how to do a meaningful experiment

Berg's proclivity for seeking out information on his own and making unilateral decisions about his future resurfaced when seeking a graduate school at which to pursue his interest in biochemistry. Once again he diligently pored over a vast collection of catalogues from schools with accredited biochemistry departments — and systematically applied to each and every one! "I must have sent out about 60 letters," he related. "That's the way I landed the two summer jobs in New Jersey. I wrote to every chemical, biological and drug and food company in the New York metropolitan area and simply waited for an attractive offer. There was no shortage of offers, so I decided to use the same strategy for entering graduate school. Some of the universities I wrote to had biochemistry departments in medical schools, others had programs in agriculture schools."[1]

In the midst of this letter writing Berg completed the coursework required for satisfying graduation requirements from Penn State. One of his assignments entailed a review of a published research paper on a topic not covered in the biochemistry course. In his preparatory readings Berg was intrigued by a collection of articles in the *Journal of Biological Chemistry* (then one of the most prestigious basic science journals in the world) that reported the impressive gains made by the recent introduction of radioisotopes to trace metabolic reactions. Rudolph Schoenheimer's treatise *The Dynamic State of Body Constituents*, a hefty book featuring isotopic labeling, which detailed how stable isotopes added to the diet can be traced to tissue lipid and protein constituents, made a particular impression. Further reading disclosed a series of papers describing the use of the newly available radioisotopes of carbon — carbon-11 and carbon-14.

From reading the biochemical literature Berg began to appreciate the utility of tracking carbon atoms from radioactively tagged biological compounds such as glucose, or intermediates of the Krebs cycle, to carbon atoms in glycogen and amino acids in proteins. He also appreciated that this technology appeared to be opening a new era in biochemistry. His appraisals revealed that many of the papers reporting these and other experiments using radioisotopes originated from the Department of Biochemistry at Western Reserve* University in Cleveland, Ohio, an institution about which he knew absolutely nothing. "I'd never heard of the place," he stated. Following a formal merger with the neighboring Case Institute of Technology in 1967 this highly regarded educational institution has carried the name Case Western Reserve University.

Cleveland became an important industrial city during the late 19th century. It was also the home of John D. Rockefeller's family.[2] Though he was born in New York, Rockefeller's Baptist family moved to Cleveland in 1853, where young John attended Cleveland Central High School and where he launched the fabulously successful Standard Oil Company during the 1860s. A man of enormous wealth, Rockefeller tithed 10 percent of his earnings to his Baptist church, beginning with his first paycheck. In due course he became one of the earliest great benefactors of medical science. In 1901, he founded the Rockefeller Institute for Medical Research — but in New York rather than Cleveland. In fact, Cleveland's educational institutions never profited directly from Rockefeller's largesse. Some say that the

---

* In 1662, Charles II gave a conveyance to Connecticut that included "all of the territory of the present state and all of the lands west of it, to the extent and breadth, from sea to sea." This gave to Connecticut, aside from the home state, the upper third of Pennsylvania, about one-third of Ohio, and parts of what has become Indiana, Illinois, Iowa, Nebraska, Colorado, Utah, Nevada, and California that was referred to as the Western Reserve. [http://homepages.rootsweb.ancestry.com/~maggieoh/csw.html]

city was not dynamic enough for his taste; others state that the city of Cleveland spent more time figuring out how to tax him than encouraging his philanthropy![3] Regardless, Case Western Reserve University is now a leading educational institution that lays claim to training and/or employing 16 Nobel laureates in Physiology or Medicine, Chemistry (including Berg), Physics and Economics. It is perhaps most celebrated as the site of the famous Michelson–Morley interferometer experiment conducted in 1887 by Albert Michelson (of Case Institute of Technology) and Edward Morley (of Western Reserve University), an experiment that measured the speed of light and simultaneously provided circumstantial evidence substantiating Einstein's special theory of relativity.[4]

In the course of preparing for the required biochemistry paper at Penn State, Berg had noted that Harland Wood, a prominent figure in research with radioisotopes, was chairman of the biochemistry department at Western Reserve. Better still, by an unanticipated stroke of luck he encountered an advertisement in a publication called *Chemical and Engineering News* offering a PhD research assistantship in that esteemed department. Berg eagerly submitted the required application in search of a graduate student position. Much to his disappointment, he was informed that all graduate student positions were filled.

****

Some of the more encouraging responses to Berg's mountain of applications to graduate schools pointedly included the caveat that he should not view acceptance to graduate school as a backdoor entrance to medical school. Their subtle religious motivation did not escape him. "I was aware of the bias about admitting Jews to medical schools and found these responses unsettling," he stated.[5]

The early 1950s were indeed a time when anti-Semitism, though not blatant, was nonetheless alive and well in the United States, and Jewish quotas for medical school admissions certainly existed in the "better" schools. Years later the renowned biochemist Arthur Kornberg (about whom more is related later), one of just two Jewish students admitted to the University of Rochester Medical School in 1937, was unhesitatingly vocal about his own perceptions of anti-Semitic attitudes. Indeed, in the final chapter of his autobiography, *For the Love of Enzymes — The Odyssey of a Biochemist,* published in 1989, Kornberg contributed a free-standing section entitled "The Virus of Anti-Semitism."

"Anti-Semitism was quietly rampant then," he told an interviewer many years later.[6] "It was true at UCLA; it was true at Stanford; it was true throughout the Midwest. I think Josh Lederberg (a friend and scientific colleague of Kornberg's who became the first chairperson of the Department of Genetics at Stanford) was the first Jewish assistant professor appointed at the University of Wisconsin — in 1947. But by that time he was already a sensational scientific star. One letter describing Lederberg that I have personally read stated: 'he is brilliant and really doesn't have all the bad features of people of his race.'" Kornberg continued: "Johns Hopkins was among the worst of these institutions. When chaired by the famous biochemist Alfred (Al) Lehninger, who wrote what is possibly the most important textbook in biochemistry and was a good friend of mine, the Department of Biochemistry at Johns Hopkins didn't have a single Jewish faculty member for 20 years — during a time when biochemistry was densely populated by Jews, either from Europe or home grown."[7]

Arthur Kornberg's antipathy to George Whipple[†] (who was also branded with a reputation for anti-Semitic proclivities) was sufficiently well known that when in 1997, Jay Stein, the former chancellor of the University of Rochester medical school arranged for Kornberg to present a seminar there, he apologized to his guest for the fact that the only auditorium large enough to seat the large crowd for his lecture bore Wipple's name!"[8]

****

On the heels of his disappointing response from Western Reserve, Berg, once again displaying his yen for independence — but sometimes for frank impatience, accepted an offer of a graduate studentship from the Department of Chemistry at Oklahoma A&M (now called Oklahoma

---

[†] George Hoyt Whipple (1878–1976) was an American physician, biomedical researcher and medical school educator and administrator. He shared the Nobel prize in 1934 with George Richards Minot and William Murphy for their discoveries concerning therapy in cases of anemia. [http://en.wikipedia.org/wiki/George_Whipple]

State University) in Stillwater, Oklahoma, a not especially distinguished academic institution situated in a town about the size of State College in Pennsylvania. "For a guy from Brooklyn to choose to go to Oklahoma A&M was pretty radical," Berg ruefully commented. "But the school was clearly interested in me. They offered to find a place for Millie and me to live in and even secured a job for Millie in a local hospital. So we agreed to move."[9]

A few weeks before the couple was scheduled to depart for Stillwater, Berg received a written communication from Western Reserve informing him that it had acquired a last-minute graduate student opening in bio-chemistry — with supporting funds! The position was his for the taking. For a while he was in a quandary. Having accepted what he considered a generous offer from Oklahoma A&M, Berg felt awkward about turning it down at the eleventh hour. However, the opportunity to join the ranks of graduate students in the famed Department of Biochemistry at Western Reserve was simply too good to be missed.

But Berg was in for a rude shock, one akin to his earlier misjudged and abortive entrance to City College in Manhattan, New York following high school. Upon arriving in Cleveland in 1948 he discovered that he had in fact applied to and been accepted as a graduate student in the Department of *Clinical* Biochemistry, which was administratively distinct from the Department of Biochemistry proper. "Since Western Reserve University was the place from which all these exciting papers that I had read came, I had naively assumed that the two departments were one and the same entity," he contritely related. "So when the offer arrived from the Department of Clinical Biochemistry I simply failed to make the distinc-tion. I thought that I was going to be in the department whose researchers I had admired — though, perhaps, in a clinical biochemistry group."[10]

Victor Myers, Chairman of the Department of Clinical Biochemistry at Western Reserve, was at one time one of the country's leading clinical chemists. Shortly after receiving his PhD in 1909, Myers joined the New York Post-Graduate Medical School and Hospital, where he successively held the positions of lecturer in chemical pathology, professor of pathology and professor of biochemistry. Following his appointment as chairman of the Department of Biochemistry at the

University of Iowa in 1924, Myers initiated the first post-graduate program for clinical chemists, a program intended primarily for those seeking positions in what eventually became variously recognized as the discipline of clinical chemistry, chemical pathology, or clinical pathology.

Myers moved to Western Reserve in 1927 as head of the "real" Department of Biochemistry, a position he retained until 1946, when Harland Wood (whose name Berg had often encountered in the scientific literature and about whom more is related later) succeeded him. The elderly Myers was politely retained at the medical school in a new (and tiny) Department of Clinical Biochemistry. Shortly before Berg's arrival two young chemists, Jack Leonards and Leonard Skeggs, joined Myers. That was the extent of the department's faculty! When Berg first entered the premises that were to be his workplace for the foreseeable future he witnessed the harsh reality that his dream of working on metabolic studies using radiotracer technology may well become the nightmare of finding himself in the wrong department!

By the time he realized his grave error there was precious little Berg could do about the situation — few alternatives existed at that late state of the game. Besides, Millie had already obtained a nursing position at one of the Western Reserve University Hospitals. Even more dispiriting, soon after his arrival in the laboratory Berg learned that Myers had assigned him the mundane task of measuring cholesterol levels in a cohort of postmortem human hearts to determine if there was any correlation with the cause of death! His brief career as a graduate student seemed poised to take a further nosedive when, about a month later, the aging Myers passed away — and the two remaining faculty members in the defunct department adopted Berg — reluctantly or otherwise!

As things turned out, Berg was far more fortunate than he might otherwise have been. Jack Leonards (who was immediately appointed interim chair of the department) and his colleague Leonard Skeggs were collaboratively working on the development of an artificial kidney for renal dialysis. Though this was a far cry from the metabolic labeling experiments being pursued in Harland Wood's laboratory just a floor away, Berg rationalized that work with the artificial kidney was nonetheless cutting edge research in clinical chemistry. So without a word of complaint (at least to anyone within earshot) he rolled up his sleeves and got to work.

The notion of renal dialysis is credited to Dutch physician and engineer Wilhelm Kolff, who, having witnessed the demise of a 22-year-old man from renal failure was inspired to invent the first functional kidney dialysis machine. Kolff fashioned a device from cellophane tubing wrapped around a cylinder that rested in a bath of cleansing fluid. Blood was tapped from an artery and, after being cleared of urea and other toxic metabolites, was returned to the venous circulation.[11] Skeggs, who had impressive engineering skills in his own right, was convinced that with Leonards' help he could improve on Kolff's efforts.

Remarkably sanguine about a situation that might have prompted serious concern (perhaps even frank panic) in less determined individuals, Berg diligently set about his assigned duties, beginning with learning how to nephrectomize dogs! While assiduously reading the relevant literature, he noted that several putative but as yet unidentified urinary proteins were alleged to have been imbued with remarkable physiological assets. One was claimed to have anticoagulant properties, another to protect against gastric ulcers, and a third putative urinary protein had been suggested to reduce blood pressure. Their extremely low concentration coupled with the enormous salt content of urine had thus far discouraged the purification of these proteins for detailed study.

Here we witness the first of many instances in which Berg squarely faced a challenge in the laboratory — and contrived an innovative solution. The enterprising idea came to him that he might be able to exploit the dialyzer to reduce the amount of salt in the urine he routinely collected in huge jugs placed in the local men's rooms, and then concentrate the dialyzed fluids by low temperature distillation to a manageable volume that he could test for biological activity. Berg located an old freezer with an intact and functional condenser and single-handedly fashioned a distillation apparatus with which he was able to effectively distill off much of the liquid and collect reasonably concentrated samples of urine to test for biological activity. "It was really a Rube Goldberg type of operation," he laughingly recalled. "But in the end I was able to detect the putative anticoagulant factor, as well as an activity that suppressed ulcer formation in rats"[12] — an impressively innovative start for a total novice in a research laboratory.

In the midst of these experiments Berg took graduate-level courses for credit in the "real" biochemistry department, including one that once again required oral presentation of relevant papers from the current literature. He elected to present a seminar on transmethylation, a biochemically important chemical reaction in which a preformed methyl group is transferred intact from one compound to another, then a topic of considerable interest and controversy in biochemical circles.

Conventional dogma held that mammals were unable to synthesize the methyl groups of methionine and choline *de novo*. They had to be supplied in the diet. But Berg uncovered hints in the literature that if one supplemented the diet of experimental animals with folic acid and vitamin B_{12}, one could do away with the methionine requirement. Indeed, Warwick Sakami, a Japanese-born professor in the Department of Biochemistry well-versed in the use of radiolabeled substrates, had detected radioactive methyl groups in methionine and choline in the livers of rats previously injected with radioactive serine and later with radioactive formic acid.

Berg delivered a polished and thoroughly researched seminar that greatly impressed the assembled faculty. Buoyed by the enthusiastic reception to his presentation, Berg approached Sakami to express his interest in determining how 1-carbon compounds are converted to methyl groups in choline and methionine. Sakami was in turn becoming increasingly impressed with Berg's keen intellect, his enthusiasm for biomedical research and his impressive knowledge of the topic at hand. Aware that the independent youngster had inadvertently landed in the essentially defunct department of clinical biochemistry, Sakami asked Berg if he was interested in transferring to the biochemistry department proper as a graduate student, where he might undertake some of the experiments he had suggested in his seminar. Sakami sang the graduate student's praises to department chairman Harland Wood, who enthusiastically reinforced Sakami's overtures. Berg was beside himself with joy. When, in later years he published an account of these events for the prestigious Feodor Lynen Lecture delivered in 1977, Berg expressed his appreciation for the opportunity to launch his scientific career in the "real" Department of Biochemistry at Western Reserve.

> Although I only moved from one floor to the next, the change altered my life: I was brought into contact with people who loved and lived for biochemistry and thereby created an environment where that spark could be nurtured in others. Enzymology, intermediary metabolism and the trials of learning how to do a meaningful experiment occupied my waking hours.[13]

Harland Goff Wood. (www.nap.edu/openbook)

Harland Wood was unable to attend Berg's 1977 Feodor Lynen lecture. But when composing a congratulatory letter to him shortly thereafter, he wrote:

> I'm very pleased that you look on the days at Reserve with pleasure and as a valuable experience. It was for me too. When I read what you're doing and able to do with DNA and RNA it's just beyond belief and makes me very proud that I had some small part in your scientific career.[14]

Though never formally a graduate student in Wood's laboratory, when writing a review article about his career as recently as 2008, Berg continued to express his appreciation for the intellectual and moral support he received from his adopted mentor.

It was my interactions with Harland Wood that provided the inspiration and set the tone for what was to become my lifelong passion. Wood's scientific exploits as a graduate student and postdoctoral fellow at Iowa State University were legendary among the biochemistry graduate students. His devotion to research and to those who shared that commitment readily showed through his outwardly gruff manner. Hanging out with the graduate students during the many late evenings when he lingered in the laboratory were the experiences I treasured most. His unremitting honesty and forthrightness in the way he practiced science provided the model we all tried to emulate. Although he never received the Nobel Prize for his important discovery of carbon dioxide utilization by organisms other than plants and photosynthetic microbes, he was widely admired and acknowledged as one of the world's leaders in biochemistry.[15]

Around the time that he was being encouraged by Sakami and Wood to apply for a transfer to the Department of Biochemistry proper, Berg was unknowingly competing with Jerry Hurwitz (who as already mentioned subsequently became Berg's close friend and scientific colleague) for a position. Indeed, when the two applications were reviewed in committee, Berg's faculty supporters let it be known that they would not support Hurwitz's candidacy unless Berg was also admitted to the program. Correspondingly, Hurwitz's supporters gleefully announced that they would not promote Berg's candidacy unless Hurwitz was admitted. In his letter to Berg mentioned above, Wood wrote:

> In your case Paul, as far as I remember, the only reticence about admitting you as a graduate student was the question of robbing Jack Leonards' graduate students. With Jerry [Hurwitz], John Muntz [a faculty member in the department] thought he was too sloppy in the lab. But we all knew John was a bit fussy about such things and he gave in.[16]

Western Reserve University thus acquired two of its most outstanding graduate students in one fell swoop. Like Berg, Hurwitz progressed to a highly distinguished academic career spent mainly at Albert Einstein College of Medicine and later at the Sloan Kettering Institute at the Memorial Sloan Kettering Cancer Center in New York. There, at the time of this writing, he continues to operate a much scaled down research laboratory. The two biochemists have remained lifelong friends.

****

This is neither the time nor place for a lengthy discourse on the advancement of the discipline of biochemistry in the 19th and 20th centuries. But a deeper appreciation of Paul Berg's scientific career begs brief consideration of the state of the field when he entered it as a graduate student in the early 1950s. In particular, readers born in the second half of the 20th century, when biochemistry and genetics began to rapidly fuse into a single intellectual endeavor called molecular biology, may appreciate some familiarity with the era that immediately preceded the "golden age" of biology, one that was primarily devoted to deciphering the many complexities of intermediary metabolism — the collective intracellular process (primarily enzymatic) that can be succinctly defined as *the sum of*

Jerry Hurwitz (*right*) teasing Millie Berg. Paul is on the *left*. (Courtesy of Paul Berg.)

all metabolic reactions between uptake of foodstuffs and formation of excretory products. Though clearly of vital importance for the preservation of growth and development of living organisms, intermediary metabolism, embracing multiple biochemical pathways (many of which were at one time absorbed by students by wrote memory), did little to kindle the type of excitement in the minds of students that molecular biology now does.

Biochemical thinking in the 19th century was dominated by the concept of "protoplasm." Max Schultze, a German histologist, coined the term, one introduced in 1861 to designate elements of the cell outside the nucleus. Schultze wrote:

> The nucleus is surrounded by the cell substance proper, a viscous "protoplasm" (a term apparently coined as a consequence of an optical illusion afforded by the light microscopes of the day), which is non-transparent on account of being filled with granules of a proteinaceous and fatty nature —[17]

The notion of a protoplasm reigned during the long barren period of "biocolloidology," a period in the mid–late 19th century when colloids were considered as "substances unable to penetrate certain membranes, that diffuse slowly, and that cannot be crystallized." Protoplasm defined Thomas Huxley's "material basis of life," a complex and mysterious protein substance that could perform all the remarkable chemical feats that cells can do — and chemists could not.[18]

By the end of the 19th century, advances in our understanding of the organic chemistry of proteins had progressed to the point that enzymes preempted Huxley's "protoplasm." The concept of "living protein" was replaced by the notion of a collection of catalytically active protein molecules, and metabolism "in block" was replaced by metabolism in steps, each catalyzed by a specific enzyme. Essentially, enzymes became for 20th century biochemistry what protoplasm was for the 19th century.[19]

The discovery of enzymes and the rapid growth of the sub-discipline of enzymology were key to the progress of cellular biochemistry. By the early 1970s about 900 such catalytic entities were documented of which a significant number were at least partially purified and a good number also crystallized. Still, the renowned German biochemist Otto Warburg, author of the famous hypothesis that bears his name (stating that "the prime cause of cancer is the replacement of the respiration of oxygen in normal body cells by the fermentation of sugar")[20] clung to the view that cellular respiration was catalyzed by iron atoms bound to an active *colloidal structure in the cell* [author's italics] that Warburg termed *atmungsferment* (respiratory enzyme).[21] Warburg blithely dismissed (at least for a time) the work of others showing that enzymes and cytochromes play important roles in biological oxidation.[22]

For a brief period in the early 1900s enzymes were regarded as a special type of biologically active colloid. But by the late 1920s their crystallization led to the appreciation that enzymes and other proteins are large macromolecules — and colloid notions in biochemistry were mercifully laid to permanent rest.[23] Indeed, by the mid-1930s, Warburg had accepted the cytochromes discovered by David Keilin in the 1920s as true respiratory enzymes, and he began to adopt the vocabulary and techniques of enzymology.[24,25]

The quarter century between 1930 and 1954 enjoined a period of extraordinarily productive classical biochemistry that identified new enzymes, naturally occurring compounds and biochemical pathways involved in multiple and diverse aspects of cellular function. These pathways embraced *cellular respiration* (including the activation of oxygen and the role of iron in biological oxidations), the *oxidative generation of biochemical energy* [specifically the energetics of glycolysis, the Krebs (citric acid) cycle and oxidative phosphorylation] and the *activation of hydrogen* (embracing electron transfer, intracellular electron carriers and the flavoprotein enzymes). This formative period also revealed details of the *biosynthesis* of glycogen, urea, amino acids, nucleic acids, fatty acids and complex natural products. Comprehending the *fixation of carbon dioxide* [the reduction of carbon dioxide in organic compounds in bacteria and plants (a primary example of the latter being photosynthesis) and later in higher organisms, including mammals], was another major milestone — and last but by no means least, progress in biochemistry incorporated the *integration and regulation of biochemical processes* by hormones, chemical messengers and various feedback loops. Little wonder that as the discipline grew, central themes and concepts in biochemistry were sometimes unified and at other times sub-divided.

The stunning progress of work on the respiratory chain (bioenergetics) that dominated biochemistry between about 1927 and 1939 was paralleled by elucidation of the cycles of intermediary metabolism, which in due course led to the concept of the biochemical unity of the cell as a system of coupled metabolic cycles with ATP as a universal medium of energy exchange. This period of classical biochemistry was dominated by some of the most famous names in the field, including giants such as Otto Warburg, Hans Krebs, Severo Ochoa, Harold Barker, Britton Chance, Feodor Lynen, Ephraim Racker, Fritz Lipmann, Carl and Gerty Cori, Arthur Kornberg — and most recently, Harland Wood at Western Reserve University. To be sure, when Berg had his momentous encounter with Warwick Sakami as a graduate student, classical biochemistry led by Harland Wood, Warwick Sakami and Merton Utter (all outstanding biochemists) dominated research in the Department of Biochemistry at Western Reserve.

As mentioned earlier, Harland Wood pioneered the use of stable non-radioactive and radioactive isotopes in intermediary metabolism. He began his career as a graduate student in Chester Werkman's laboratory at Iowa State University. Though it was well established that plants and a few specialized bacteria fix carbon dioxide, his graduate studies led Wood to the initially controversial notion that *all* living organisms do so. He spent the next half-century elucidating the intricacies of carbon dioxide fixation in higher organisms and was active in the laboratory until a few days before his death in the fall of 1991. In a tribute written shortly after he died, Wood's faculty colleagues David Goldthwait and Richard Hanson noted:

> At the time of his death at age 84 Wood held three research grants from the NIH, oversaw a working group of 15 associates, and was writing nine manuscripts. At the last meeting of the American Society of Biochemistry and Molecular Biology that he attended, he had 12 posters on display and was present to discuss results related to each and every one of them. In the 14 years between his 70th birthday and his death, Wood published 96 papers, all in well-respected journals.[26]

In 1942, Wood received the prestigious Eli Lily Award from the Society of American Microbiologists. The award came with much-needed cash that

Wood and his wife used to purchase a home in Ames, Iowa. Upon hearing about this transaction Chester Werkman irately asked Wood, "Do you think you can stay here forever?" This was apparently the last straw in Wood's unrelentingly contentious relationship with his mentor (who, in all the years they worked together never rewarded Wood with a position higher than Research Associate — a staff rather than a faculty position), and precipitated a permanent break in their relationship. Wood obtained a faculty position in the Department of Physiological Chemistry at the University of Minnesota, where he remained until 1946 when he was recruited to take the reins in the Department of Biochemistry at Western Reserve University.[28]

Berg had the good fortune of initiating his research career at a time when radioactive isotopes became favored over stable forms for all sorts of metabolic labeling studies. "Radioactivity made this kind of experimentation much simpler than using stable isotopes," he commented. "With stable isotopes one had to have access to an elaborate mass spectrometer (Wood built his own) and had to recover large amounts of material. Whereas with radioactivity one could literally use trace quantities."[29]

This sort of research required more than a smattering of expertise in organic synthesis. Berg recalls a time when he set about synthesizing radioactive methyl alcohol from 10 millicuries of barium carbonate — a huge amount of radioactivity by any standard. He set up an organic synthesis array in which the starting material was at one end and the gaseous products had to pass through various kinds of bubblers before being collected in a reservoir. (He built the apparatus himself.) When he completed the experiment he was thoroughly dismayed to discover that he had no radioactive product! One of the glass bubblers had a pinhole in it and gas had escaped through the hole. "I had lost ten millicuries of ^{14}C — for which I was both ridiculed and the recipient of a lot of flak," Berg sheepishly related.[30] Remarkably, he failed to win the infamous horse's-ass award[‡] at

---

[‡] A long-established tradition in the department of biochemistry, Christmas parties included (among other hijinks) stiff competition for this "coveted" award, which featured the plump rear end of a horse mounted on a special plaque bearing the "proud winner's" name, which he/she was required to keep on his/her desk for an entire year. When (many years ago) the author was a postdoctoral fellow in David Goldthwait's laboratory in the Western Reserve biochemistry department, a graduate student in that

the next annual biochemistry department Christmas celebration for this "outstanding failure."

Once settled in Sakami's laboratory, Berg pursued the origins of newly synthesized methyl groups. Working independently in the main, but always under the watchful eyes of his formal mentor Warwick Sakami and his informal mentor Harland Wood, Berg established that methionine was indeed the first compound to acquire newly synthesized methyl groups and that these were transferred intact to other methyl-containing metabolites. In preliminary efforts to explain the mechanism of the conversion of 1-carbon compounds to methyl groups, Berg also established a cell-free system that supported the robust conversion of formate and also formaldehyde to methionine methyl groups in the presence of homocysteine, and obtained preliminary evidence for a role for folic acid derivatives and Vitamin B12 in this conversion.

In September 1952, Berg completed a PhD thesis entitled *A Study Of the Conversion of Formate To Labile Methyl Groups*. These findings were made public in two articles, one of which consumed a hefty 17 pages in the distinguished *Journal of Biological Chemistry*.[31] Though gratefully acknowledged for advice and guidance, Berg's thesis advisor Warwick Sakami did not append his name to either of these papers, a generous gesture presumably offered in acknowledgement of his graduate student's capacity for independent experimental and intellectual contributions — and one not frequently encountered in academia!

Berg's intellectual acuity began to garner attention outside Western Reserve. Memorably, after presenting his work at an annual meeting of the Federation of American Societies for Experimental Biology in Atlantic City, New Jersey, he was engaged in a debate by Vincent du Vigneaud, professor of biochemistry at Cornell University. One of the doyens of transmethylation and a future Nobel Laureate in chemistry, du Vigneaud had been protecting the principle that some source of labile methyl groups was absolutely required in an animal's diet. "And here I was, a young graduate student, not only stating that they were not required, but

---

laboratory won the award for following a directive to get rid of a large number of old scintillation vials by attempting to flush them down the toilet – thoroughly clogging it!

also showing how they could be synthesized in the body," Berg related.[32] Not too much later du Vigneaud wrote to Harland Wood expressing his admiration for Berg and inquiring whether the rising young star might be interested in a faculty position at Cornell. He was astounded to learn that Berg was a mere graduate student!

Wood's vibrant department was a frequent stopover for many of the leading biochemists of the day, and Berg delighted in putting faces to the many famous names he had heard and read about, not to mention opportunities to engage them about science — and the possibility of doing postdoctoral work with one or more of them. He was especially pleased to learn that like himself, Arthur Kornberg had attended Abraham Lincoln High School in Brooklyn New York.

While at Western Reserve, Berg also met the famous Danish biochemist Herman Kalckar. Born in Copenhagen in 1908 into a Jewish-Danish family, Kalckar obtained a degree in medicine at the University of Copenhagen in 1933 and immediately became a candidate for the PhD degree in the Department of Physiology. In the early 1930s, unable and perhaps also unwilling to work in Nazi Germany any longer, Fritz Lipmann moved to Copenhagen, where he became one of Kalckar's mentors and a lifelong friend.[33]

Classic studies by Pasteur had demonstrated that aerobic metabolism of glucose by yeast is vastly more energy efficient than the anaerobic process. Between 1937 and 1939, Kalckar addressed the pressing question: "How is energy captured by the oxidation of sugars and other foodstuffs linked to the reduction of molecular oxygen?" His experiments led him to the demonstration that cell-free extracts of kidney cortex catalyze oxidative phosphorylation — the formation of ATP in reactions strictly dependent on the reduction of oxygen and independent of glycolysis. This work gained him substantial attention in biochemical circles. Berg was intrigued by Kalckar's work on oxidative phosphorylation — and, especially by the notion of spending a year in Copenhagen as a postdoctoral fellow — an idea that soon matured to reality.

****

An attic room in a private home in the suburb of Cleveland Heights was all that Paul and Millie were initially able to locate in order to get settled. Though the house was in a relatively pleasant and safe part of town, the bizarre behavior of their landlord soon made finding an alternative an imperative! The proceeds of the GI Bill and the modest graduate student stipend Berg expected to receive would, they knew, put a dent in the extravagance of the living quarters they would have to settle for. Unfamiliar as they were with living locations in Cleveland, they eventually located a three-room apartment in a dismal neighborhood of East Cleveland — an area of town in which to be wary at any time, and one most certainly not to be explored at night! Many hours of repair, scrubbing and decoration finally made the place habitable. The couple still did not own an automobile, so they had to rely on local transportation for commuting to and from the University Hospital and Medical School.

Millie began work on the Obstetrics Service at the University Hospital. Her hours were initially 7 am to 7 pm six days a week, a brutal schedule that necessitated rising well before the sun was up. And since it was usually dark when she returned in the evenings, winter spelled many sunless days for her. After about six to nine months, the Bergs made their first investment in a car — a used but reliable enough Chevrolet. While that acquisition reduced some of the misery of the dreary daily commute from East Cleveland to the university campus, the Bergs (after paying a $100 bribe) were able to secure a more convenient and comfortable abode in a building very close to where they both worked. Happily, their friends Jerry Hurwitz and his wife Muriel lived in the same building.

Like many students of the day eager to graduate as soon as possible, Berg often worked in the laboratory at night, a schedule that with Millie's daylight work hours took its toll on their personal lives. But in due course she was able to acquire an alternative schedule of 6 pm to midnight, affording the couple occasional time together during the day. This restructured work schedule allowed Berg to remain at the bench until close to midnight, a convenient time to collect Millie from the hospital for the short walk to their nearby apartment. Still, Berg recounts: "[t]he four years in Cleveland were difficult for both of us as I spent long hours in the laboratory and Millie worked a six to midnight shift, which didn't leave much time for relaxation outside of the weekends.[33]" But living in close proximity to the hospital had other advantages. Severance Hall, the home of the Cleveland symphony orchestra under the direction of famed conductor George Szell, was within easy walking distance. And when one bitter winter produced a massive snowfall that paralyzed all public conveyances for a spell, the Bergs were among the few who made it to work.

Cleveland, Ohio, was infamous for the fact that the Cuyahoga river (which flows through the middle of the city) was so polluted with flammable material from indiscriminate dumping of products of the gigantic Cleveland steel industry that it once caught alight — making Cleveland the laughing stock of the country. The winters were notably dreary and the winds off the lake, even without snow, made spending time outdoors off-putting — except for attending the Cleveland Browns Sunday football games! Both being ardent sports fans, Berg and Hurwitz acquired season tickets for home games. Millie and Muriel, who usually tagged along, had little choice but to endure sub-zero temperatures, all the while uncomfortably surrounded by more boisterous football fans. But Cleveland did offer more cultural pursuits. In addition to the symphony orchestra, the city boasted an equally notable art museum and an active and highly regarded theater company. Additionally, the nearby heavily Italian section (affectionately referred to as "Little Italy") provided enticingly satisfying restaurants with prices suitable for graduate students. Deeply immersed in their work, neither Paul nor Millie had much time for socializing. But free evenings and weekends were often spent in the company of Jerry and Muriel Hurwitz.

Notes

1. ECF interview with Paul Berg, April 2011.
2. http://en.wikipedia.org/wiki/John_D.Rockefeller
3. http://cleveland.about.com/od/famousclevelanders/p/rockfeller.htm.
4. http://en.wikipedia.org/wiki/Case_Western_Reserve_University#Nobel_Laureates
5. ECF Interview with Paul Berg, April 2011.
6. AK Interview by Sally Hughes, Regional Oral History Office of the Bancroft Library at the University of California, Berkeley, 1998.

7. *Ibid.*
8. *Ibid.*
9. ECF interview with Paul Berg, April 2011.
10. *Ibid.*
11. http://www.engology.com/eng5kofff.html
12. ECF Interview with Paul Berg, April 2011.
13. Berg P. (1977) The Eighth Feodor Lynen Lecture — Biochemical Pastimes-and Future Times. *Molecular Cloning of Recombinant DNA, Miami Winter Symposium*, Vol. 13. Academic Press.
14. Letter from Harland Wood to Berg, dated February 7, 1977. Reproduced with permission from the Department of Special Collections, Stanford University Libraries.
15 Berg P. (2008) Moments of Discovery. *Annu Rev Biochem* **77**: 15–44.
16. Letter from Harland Wood to Berg, dated February 7, 1977. Reproduced with permission from the Department of Special Collections, Stanford University Libraries.
17. Florkin M, Stotz E. (1972) *A History of Biochemistry*, Comprehensive Biochemistry 30: p. 295, Elsevier.
18. Kohler RE. (1975) The History of Biochemistry: A Survey. *J Hist Biol* **8**: 275–318.
19. *Ibid.*
20. http://en.wikipedia.org/wiki/Otto_Heinrich_Warburg
21. Kohler RE. (1975) The History of Biochemistry: A Survey. *J Hist Biol* **8**: 275–318.
22. *Ibid.*
23. Fruton J. (1990) *Contrasts in Scientific Style — Research Groups in the Chemical and Biological Sciences*. Amer Phil Soc, Philadelphia, p. 209.
24. Deichmann U. (2007) Molecular Versus Colloidal: Controversies In Biology and Biochemistry, 1900–1940. *Bull Hist Chem* **32**: 105.
25. Kohler RE. (1975) The History of Biochemistry: A Survey. *J Hist Biol* **8**: 275–318.
26. Goldthwait DA, Hanson RW. (1996) *Harland Goff Wood. 1907–1991. A Biographical Memoir*. National Academies Press, Washington DC, pp. 395–428.
27. Berg P. (2008) Moments of Discovery. *Annu Rev Biochem* **77**: 15–44.
28. Singleton R. (1997) Harland Goff Wood, An American Biochemist. In: Semenza G, Jaenicke R (eds), *Comprehensive Biochemistry: History of Biochemistry*. Vol. 40. Amsterdam, Elsevier.

29. ECF interview with Paul Berg, April 2011.
30. *Ibid.*
31. Berg P. (1951) *J Biol Chem* **190**: 31–38; Berg P. (1953) *J Biol Chem* **205**: 145–62
32. ECF interview with Paul Berg, April 2011.
33. http://www.nap.edu/html/biomems/hkalckar.htm
33A. Paul Berg, *Millie and I*, personal communication.

CHAPTER FIVE

Copenhagen

The rabbit was not an enthusiastic swimmer

As completion of his research toward a PhD degree approached, Berg contemplated where he might continue his training as a postdoctoral fellow. Harland Wood entertained his own ambitions for his protégé and had in fact covertly arranged for Berg to join Carl and Gerty Cori's laboratory at Washington University — then considered one of the finest biochemistry laboratories in the world.

Among the most celebrated biochemists of their era, Carl Cori and Gerty Radnitz[1] hailed from Austrian families that had lived in Prague for generations. The couple met at university, where they received medical degrees in 1931 — and subsequently married. The many difficulties of life in post-World War II Europe, complicated by Gerty's Jewish heritage (not to mention the fact that she was a woman aspiring to become a scientist in an almost exclusively male-dominated profession) prompted the couple to think seriously about emigrating to the United States. No offers followed an initial search for research appointments, so Carl accepted a position with Otto Loewi in the Pharmacology Department at the University of Graz in Vienna. Notwithstanding the undeniable satisfaction of working with Loewi (a celebrated biochemist who discovered acetylcholine and earned a Nobel Prize in Physiology or Medicine in 1936), Carl found the atmosphere at Graz increasingly uninviting. Living conditions were generally poor and he was irked by the requirement to prove his Aryan descent in order to obtain employment at the university. Once again the couple turned their attention to the US, this time with a promising outcome. In 1922, job offers came for both Coris from the State Institute for the Study of Malignant Disease (later the Roswell Park Cancer Institute) in Buffalo, New York.

In 1931, Carl Cori was offered the chairmanship of the Department of Pharmacology at Washington University School of Medicine in St. Louis. His only previous academic experience was as an adjunct assistant professor at the University of Buffalo for one year. Gerty was offered a research position in the same department — at a salary a fraction that of her husband's! In 1942, Carl Cori received a joint appointment as professor of biochemistry. Four years later he became head of that department.

The Coris' most notable contributions to science embraced a series of discoveries that elucidated the pathway of glycogen breakdown in animal cells and the enzymatic basis of its regulation. Their groundbreaking research into the enzyme-catalyzed chemical reactions of carbohydrate metabolism elevated the medical school at Washington University to the front ranks and won the Coris Nobel Prizes in Physiology or Medicine in 1947. Sadly, the excitement of receiving Nobels was tainted by the discovery that Gerty had developed myelosclerosis, an incurable and fatal illness. She died in 1957.[2]

The Coris left a rich legacy to the world of biochemistry that notably included training successive generations of outstanding biochemists. No fewer than six future Nobel Laureates (Christian de Duve, Arthur Kornberg, Edwin G. Krebs, Luis F. Leloir, Severo Ochoa and Earl W. Sutherland) passed through their laboratory at Washington University. It was in the Cori laboratory that Harland Wood wished to see his young protégé continue his career.

Although Berg (who the reader presumably now readily appreciates possesses a distinctively autonomous personality) was well aware of the scientific fame and accomplishments of the Cori couple, he was not at all keen to move to St. Louis, a city about which he had heard nothing good! Culturally more Midwestern in character than Cleveland, St. Louis was also known to be politically conservative — a place where African-Americans were only recently freed of the necessity of having to sit in the back of the bus in the mid-1950s. Nor did living through the intensely hot and muggy summer months enhance its appeal to Berg.

After hearing of Herman Kalckar's and Arthur Kornberg's research during their visits to Western Reserve University, Berg decided to pursue one year of postdoctoral training in Kalckar's laboratory in Copenhagen, Denmark, followed by a second year with Arthur Kornberg at the National Institutes of Health in Bethesda, Maryland. Except for his limited exposure while in the US Navy, Berg had never set foot outside the United States, and he and Millie were eagerly anticipating the year in Copenhagen. Wood was both astonished and piqued that anyone would turn down an opportunity to work side by side with the celebrated Coris. But despite his insistent urging, Berg (who by then believed that his revered mentor may well have revised his glowing opinion of him) was adamant about not moving to St. Louis.

In the early 1950s, overseas travel was by no means the commonplace experience it is now — and most certainly nothing to be blasé about. Recall that commercial jet air travel was only inaugurated in 1952 — and was prohibitively expensive by the standards of the time. Television (only in black and white of course) was then just emerging as a common household commodity, ballpoint pens were still novel and recorded music came on discs called "records," typically played on small portable gramophones that required manual winding. So Paul and Millie were appropriately excited at the prospect of their first visit to historic Europe — and imbibing a heavy dose of "real culture."

The couple left New York by ship in October 1952 bound for England. Millie relished the freedom that had eluded her during her four years of intensive nursing and was perfectly content to spent entire days reclining on a deck chair, reading, taking in the sun and napping, waking only when the ship stewards offered morning coffee or afternoon tea, events always accompanied by mouth-watering snacks.

Paul, on the other hand, discovered to his dismay that his vulnerability to seasickness was not limited to small submarine chasers in stormy waters and spent much of the first half of the voyage suffering in his tiny cabin. "I might have starved had Millie not plied me with light food she brought to the cabin," he commented. "Having spent time in the Navy, my inability to manage even the slightest rocking of the ship was a particular embarrassment."[3]

When Berg recovered, the voyage was lightened by socializing with fellow travelers, one of whom, a young Danish pilot, gallantly offered to help the American couple get their tongues around a few words in Danish, a language they eventually concluded only Danes could master! Berg was in fact reminded of a comment from a Danish humorist that Danish pronunciation was evocative of a throat disease! The final night on board was spent enjoying the traditional Captain's Dinner — which only ended as the English coastline emerged with the rising sun.

The Bergs visited historic sites in London, Oxford and Paris before boarding a plane to Copenhagen, where their vacation was extended a while longer before Paul settled down to work. Both he and Millie were deeply impressed by the gracious cordiality and informality of the Danes, many of whom were excited to meet Americans for the first time. A particularly memorable experience unfolded shortly after their arrival when they were perusing a menu (in Danish of course) posted outside a modest looking Copenhagen restaurant. Berg, who documented his post-doctoral year in a lengthy unpublished essay dubbed *A Year Abroad (1952–3)*, picks up the story:

> A relatively young man standing alongside said something to us in Danish; we couldn't make out if it was a recommendation or a caution. Millie responded in her only recently learned phrase, saying that we didn't understand Danish, whereupon he asked if we were French. When we confessed that we were Americans, he addressed us in perfect English informing us that the restaurant was under new management and the food was excellent. Being somewhat cautious about advice from a stranger we hesitated, but eventually entered what turned out to be a very attractive restaurant. Before we were seated, Eric Eng formally introduced himself and invited us to join him and his fiancée Rutte for dinner. Apparently well acquainted with the chef, Eric insisted on ordering dinner for us. The meal was replete with renowned Danish schnapps, magnificent Carlsberg beer and a course of shrimp smorrebrod followed by a wonderful Weinerschnitzel, all the while accompanied by most congenial conversation.
>
> Eric insisted on showing us "his Copenhagen", whereupon we continued the evening with successive visits to nightclubs, frequently

sampling the Danish cherry liquor called Cherry Heering.* At one club he serenaded us with piano renditions of Gershwin and other popular songs and then took us to a most elegant club where we danced to a full orchestra and "skoaled" more Cherry Heering. Finally, we invited Eric and Rutte to join us on the Codan Hotel roof garden where brandy and coffee ended the long night. A most unusual and unforgettable introduction to Danish hospitality and the beginning of a year-long warm friendship.[4]

Denmark was still recovering from the heavy toll of World War II. Food and other simple pleasures were rationed and when available for purchase were prohibitively expensive. When the Bergs first visited the famous Georg Jensen silver store in Copenhagen they were surprised to find a substantial crowd intently peering at the window display. It took a

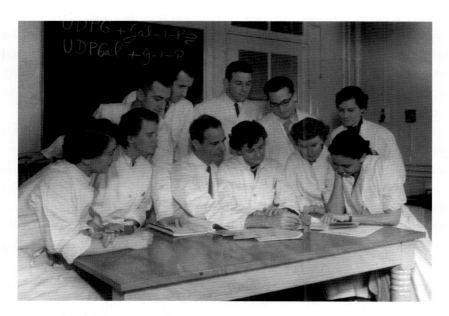

Herman Kalckar's laboratory staff in Copenhgen circa 1952. Berg is third from the right in the back row.

---

*Proprietary Danish cherry liqueur with brandy base has been produced since 1818 under various names such as "Heering," "Peter Heering," and "Cherry Heering." Highly valued in cocktails, cherry heering is considered the best of the dark cherry liqueurs. [http://www.cocktaildb.com/ingr_detail?id=149]

few moments to realize that the appeal was not the silverware — but the juicy looking peaches that adorned many of the silver bowls.

The annual stipend ($3,600) that came with Berg's American Cancer Society fellowship was a handsome salary by Danish standards. Indeed, a Danish friend jokingly suggested that they were probably earning more than the King of Denmark! The couple could afford to rent a charming two-story villa in a small fishing village (Taarbaek) on the outskirts of Copenhagen, directly opposite a huge deer park that once accommodated sport hunting by Danish royalty. "The daily commute to and from the lab was one of the treasured times when I could think about experiments to be done that day or to review the data from those done during the day," Berg related.[5]

Berg was fortunate to find Kalckar in residence. Jim Watson had preceded him by a year and saw very little of his intended mentor, who was then gallivanting around Europe with his new love, Barbara Wright, an American postdoctoral fellow who he subsequently married. Watson was sufficiently disenchanted by Kalckar's absence that he orchestrated a move to Ole Maaløe's laboratory in Copenhagen, and later in the year to Cambridge University — where he teamed up with Francis Crick.

Despite years spent in the United States, Kalckar was renowned for his almost unintelligible English. "I slowly learned to understand him," Berg commented, "and when I couldn't I just kept plugging away — trying my best to avoid looking overtly embarrassed. I found that the easiest way to communicate with him was to use a blackboard. Sometimes we resorted to emulating experiments pictorially! Eventually I was not only able to follow his diction, but began to anticipate his thinking."[6]

Time away from the laboratory was cheerfully dedicated to visiting museums and art galleries and taking in other tourist delights that Copenhagen offered. The city's opera and ballet companies were well known and their museums held art treasures the couple had only seen in photographs. Millie soon learned enough Danish to be comfortable with the shopkeepers in the village in which they resided, and with some of their Danish friends who spoke no English. The Bergs learned to love and relish Danish foods, particularly open-faced sandwiches, smørrebørd, and renowned Danish beers — *Carlsberg* and *Tuborg*. During their stay in Denmark the Bergs traveled extensively to ski in Norway, visiting the old university in Uppsala, Sweden, and also throughout Germany, Italy and Switzerland before returning to the United States for Paul's second post-doctoral year.

One evening Kalckar invited Berg and another foreign postdoctoral fellow to hear him address the Danish Royal Academy of Sciences. Another speaker presented an analysis of a Danish pornographic poem! Not wanting to miss any of the anticipated juicy report, the elderly members of the Academy adjusted their earphones to catch every nuance. Naturally Berg and his colleague couldn't understand a word. When Kalckar rose to present some of the work going on in his laboratory, all the elderly attendees removed their earphones. Even when speaking Danish, Kalckar was difficult to interpret!

The Kalckar laboratory offered a congenial and interactive environment. "Kalckar convened lab meetings at the end of each day," Berg recalled, "and if something substantial had transpired in the laboratory that day he would bring out a bottle of Cherry Heering and we'd all drink a toast — sometimes multiple toasts — to which ever person(s) was being congratulated that day."[7]

"Herman was something of a dreamer," Berg continued, "often seeking novel explanations for paradoxical observations, some of which surfaced in conversation during afternoon tea." A productive (and amusing) experimental outcome of one such conversation ultimately led Berg and Bill Joklik (an American colleague in the Kalckar laboratory) to the discovery of a novel enzyme.

In the course of experiments that required quantities of radiolabeled ATP (or ITP) far greater than the laboratory could afford to purchase, Berg and Joklik decided to isolate their own ^{32}P-ATP. First they encouraged (more accurately forced) a rabbit to swim to exhaustion in a large bathtub in order to deplete its muscle ATP. After the injection of a healthy dose of radiolabeled inorganic phosphate, the rabbit was allowed to rest, ostensibly to replenish its ATP stores. The animal was not an enthusiastic swimmer, nearly drowning in the process! But Berg and Joklik recovered sufficient ^{32}P-ATP from the rabbit's muscle tissue for their experiments.

These experiments providentially led to the serendipitous discovery that the terminal phosphate of ATP can be transferred to IDP and from ITP to ADP. Pursuing this unanticipated observation Berg and Joklik discovered a novel enzyme that catalyzed a transphosphorylation reaction, which utilized ATP to phosphorylate the four ribo- and deoxyribo-nucleoside diphosphates to their respective triphosphates. They christened the new enzyme nucleoside diphosphokinase. These observations yielded a paper in *Nature* entitled "Transphosphorylation Between Nucleoside Polyphosphates" and another called "Enzymatic Phosphorylation of Nucleoside Diphosphates," which consumed 15 pages of the *Journal of Biological Chemistry*. Berg's career as a biochemist was clearly in the ascendancy.

Notes

1. http://beckerexhibits.wustl.edu/mig/bios/coric.html.
2. *Ibid.*
3. Interview with Paul Berg, April 2011.
4. Paul Berg, *A Year Abroad*, personal communication.
5. Interview with Paul Berg, April 2011.
6. *Ibid.*
7. *Ibid.*

PART II

CHAPTER SIX

Washington University, St. Louis

Simply stated, I discovered a new kind of reaction

While in Copenhagen, Berg received a letter from Arthur Kornberg, with whom he had arranged to spend a second postdoctoral year at the NIH following his stay in Copenhagen. To his surprise he was informed that Kornberg had accepted the chairmanship of the Department of Microbiology at Washington University — in St. Louis, Missouri, of all places! The prospect of living there had to be faced once again! But this time there was no backing off. For one thing, Berg viewed Kornberg as one of the brightest stars in the biochemistry firmament. Then too, Kornberg had been more assertive than Cori in his expressed interest in having Berg join his laboratory. "Cori would have taken me at Harland Wood's behest," Berg stated. "But I had a sense that Arthur really wanted me to join his lab."[1]

Looking to the future, Berg was intent on pursuing an academic career as a biochemist. With the news that Kornberg would soon be building a new basic science department at Washington University, he surmised that he might be in line for a junior faculty position at some point. "At least he'd be able to give me a thorough looking over," Berg declared.[2]

Though Berg had no choice but to come to terms with his earlier prejudices about St. Louis, the physical space that Kornberg inherited when he arrived at Washington University was disappointing. The building that housed the Department of Microbiology was once a medical clinic for indigents and was in awful physical condition. A rickety elevator took one to the floor where the department was housed. Oliver Lowry, then head of the Pharmacology Department at Washington University, loved to tinker as a handy man in his spare time. Lowry reassured Kornberg that he would personally repaint the laboratories and offices and redo the electrical wiring. He never did so of course!

Ultimately Kornberg threatened the leadership of the medical school, saying that he was going to leave unless promised laboratory renovations materialized soon! By the time of Berg's arrival, several laboratories had been redone to a usable state, and others followed. Quickly settling into his new postdoctoral position, Berg soon made another important discovery in the enzymology of intermediary metabolism — one that ultimately provided his entree to the world of nucleic acids, an arena in which he would remain active for the remainder of his academic career.

This seminal contribution from Berg is an object lesson in how science sometimes fortuitously progresses from chance observation to discovery. While in Copenhagen, Berg was struck by a brief paper in the *Journal of the American Chemical Society* entitled "On the Enzymatic Mechanism of Coenzyme A Acetylation with Adenosine Triphosphate and Acetate."[3] The paper was a collaborative effort by two research groups headed by the celebrated European biochemists Fritz Lipmann and Feodor Lynen, which, according to Berg, "got lots of discussion at our tea times."

Lipmann was one of the many Jewish scientists who left Germany for the United States when the Nazis came to power. Lynen (who was not a Jew) remained in Germany. Not surprisingly, questions, some possibly articulated publically, were raised about Lynen's political bent during WWII.* In 2004, one of Lynen's former graduate students,

---

* Lynen never publicly defended his decision to remain in Nazi Germany. But by all accounts he never remotely considered joining the Nazi party, and was fortuitously exempt from service from both the German military structure and all Nazi paramilitary organizations, perhaps the only benefits of a permanently damaged knee following a skiing accident in 1932 which left him with a pronounced limp. Regardless of the fact that he never joined the National Socialist German Workers' Party, Lynen was appointed Docent of Chemistry in 1942 — though with the qualification "only with concern." [Everything Now Seemed So Simple to Me — : Feodor Lynen (1911–1970), a Hero of Biochemistry, *Angew. Chem Int. Ed.*50: 1158011584, 2011.]

Hildegard Hamm-Brucher, who became a prominent liberal politician in the Federal Republic of Germany in later years, remembered her thesis advisor thus:

> He neither endangered his life by conspirative activities nor by death-condemning conduct, but he has always proven himself as an example of integrity and civil courage and endeavored to preserve the chemical institute as an oasis of decency.[4]

In later years, the Alexander von Humboldt Foundation established a research stipend in Lynen's name. Additionally, the German Biochemical Society annually identifies a distinguished scientist as the Feodor Lynen Lecturer, a distinction that, as mentioned earlier, was awarded to Berg and earlier to Harland Wood.

In the period immediately following the war, their European and American colleagues spurned scientists who remained in Germany during the war years. Indeed, only four German biochemists were invited to attend the First International Congress of Biochemistry convened in Cambridge, England, in July 1949. Lynen was one of them.[5] In 1953, Lipmann shared the Nobel Prize in Physiology or Medicine with Hans Krebs — of Krebs cycle fame. Nine years later, Lynen won the coveted

Feodor Lynen (1911–1979). (www.nobleprize.org)

award with Konrad Bloch, another outstanding American cum German biochemist.

Armed with degrees in medicine and chemistry, Fritz Lipmann spent his entire academic career as a biochemist. He emphasized the significance of phosphates and the transfer of phosphate groups in bringing about energy transformation in cells. He also introduced the terms "high energy phosphate bond" and "phosphate transfer potential" and he pioneered the famous squiggle (~) denoting these bonds. Kalckar too published a lengthy review article on biological energy that year. Both articles drew attention to the thermodynamic instability of molecules said to possess "high energy."

Lipmann was a pioneer in linking descriptive biochemistry with energy transformations in cells, and his focus on the role of phosphates in energy metabolism revolutionized biochemical thinking in many ways. It was not for this work, however, that Lipmann shared the Nobel Prize with Krebs, but rather for his discovery in 1947 of coenzyme A and its importance in intermediary metabolism.

While working on the role of phosphate in cellular metabolism, Lipmann discovered a heat-stable factor that acted as a carrier of acetyl (CH3CO-) groups. He eventually isolated and identified what he called acetyl-coenzyme A (often referred to as "cofactor A," or "CoA") and showed that it contains vitamin B_2. He additionally demonstrated that a two-carbon compound known to join with oxaloacetic acid to generate citric acid during the Krebs cycle was in fact acetyl-CoA.

Though largely isolated in Germany during the war years, Lynen was investigating so-called "active acetate." In 1950, the coenzyme was shown to have wider applications than the Krebs cycle based on Lynen's discovery that it plays a key role in the metabolism of fats.[6] When, at war's end, scientific contact with the rest of the world was re-established, Lynen boned up on the many recent advances in this area of biochemical research and learned of the exciting observation that pantothenic acid (also known as vitamin B_5) containing coenzyme A participates in the formation of "activated acetic acid."

Acetyl-coA is present in both prokaryotes and in more highly evolved organisms. In the early 1950s it was established that in bacteria this compound is synthesized via two reversible enzymatic reactions.

1. ATP + acetate ↔ acetyl phosphate + ADP (catalyzed by an enzyme called acetokinase)
2. Acetyl phosphate + CoA ↔ acetyl-CoA + Pi (catalyzed by a transacetylase)

In contrast, the prevailing dogma held that in yeast and higher eukaryotes acetyl-CoA was synthesized in a single reaction:

3. ATP + acetate + CoA ↔ acetyl-CoA + AMP + PPi

In 1953, Lipmann, Lynen and their colleagues published the paper referred to above, in which they proposed an alternative reaction scheme in higher organisms involving *three* steps:

1. Enzyme + ATP ↔ Enzyme-AMP + PPi
2. Enzyme-AMP + CoA ↔ Enzyme-CoA + AMP
3. Enzyme-CoA + acetate ↔ Enzyme + acetyl-CoA

This formulation was based on their observations that following incubation of an enzyme preparation from yeast with radioactively labeled pyrophosphate (PPi) and ATP, the radiolabeled phosphates ended up in ATP. Similarly, the enzyme preparation promoted the exchange of radiolabeled acetate with the acetyl group of acetyl-CoA. Lippmann and Lynen confidently announced that they had solved the mechanism for acetyl-CoA formation in eukaryotes. Not too long after his arrival in St. Louis, Berg convincingly proved otherwise!

When he arrived at Washington University in the fall of 1953, Berg was told that of four postdoctoral fellows who had been accepted to join the Kornberg laboratory in Bethesda, Maryland, he was the only one who had agreed to come to St. Louis! He might even have mentioned to some that he came close to being the fourth naysayer! Perhaps the idea of living in St. Louis was an anathema to others besides Berg! One of the other four who refused the invitation to come to St. Louis was named Edward Korn. In his autobiography *For the Love of Enzymes — The Odyssey of a Biochemist*, published in 1989, Kornberg could not resist contemplating that "had Korn eventually joined me, as Berg did, we might have produced a paper by Korn, Berg, and Kornberg!"[7]

The Kornberg laboratory was then immersed in two major research arenas: deciphering the biosynthesis of phospholipids, and the synthesis and degradation of pyrimidines. Kornberg invited Berg to pursue aspects of either of these projects. But struck as he was by Lipmann's and Lynen's postulated involvement of AMP in the synthesis of acetyl-CoA, Berg wondered whether cells might contain other enzyme-bound nucleoside monophosphates (such as GMP-, CMP- or TMP-enzyme complexes) that might serve as precursors for nucleic acid biosynthesis — though he concedes that he may have first acquired this notion when he read the Lipmann, Lynen, *et al.* paper just referred to, which concluded with the following prescient sentence:

> It seems rather attractive — to speculate that the here-observed formation of an enzyme mononucleotide may well foreshadow this as a general biosynthetic mechanism involved, for instance, possibly in nucleic acid synthesis.[8]

This view was restated in Lipmann's Nobel Lecture presented in December 1953, when in the last sentence of the published transcript of the lecture he spoke "very tentatively of the first clues that may fore-shadow extension of his concept to proteins and nucleic acids."[9]

Kornberg witnessed a classic example of his young colleague's inde-pendent and fearless streak when Berg politely informed him that rather than pursuing work on lipid biosynthesis or the synthesis and degradation of pyrimidines, he wished to pursue the isolation of enzyme-bound nucle-otidyl complexes and explore the possibility that comparable enzyme complexes with other nucleotides may exist. To his surprise, Kornberg relented without argument. "Arthur had a history of directing work in his laboratory," Berg stated. "He more or less decided what people could and could not do. So it was unusual for him to allow me to venture off on my own. On the other hand, perhaps he thought well enough of me that he decided to give me a green light to explore something other than what was going on in the lab at the time."[10] This being said, Kornberg predicted that Berg's efforts would turn out to be a waste of his time and energy, a predic-tion in large part motivated by his skepticism about Lynen's and Lipmann's observations and claims — about which he had his own interpretation.

"Don't waste clean thoughts on dirty enzymes" is a famous Kornberg aphorism* admonishing the interpretation of experiments that include enzymes that may be contaminated by impurities — especially other enzymes. As a dutiful adherent to this oft repeated golden rule Berg began his quest by purifying acetyl-CoA synthetase from extracts of brewer's yeast, using the formation of acetyl-CoA as an assay.[11,12]

Within a relatively short time he obtained a considerably purified enzyme preparation, but much to his surprise neither of the two kinds of exchange reactions reported by Lipmann, Lynen and their colleagues were detectable. Bent on determining the requisites for reconstituting ATP-PPi exchange, Berg observed an absolute requirement for acetate. He inferred that ATP reacted with acetate to produce acetyl adenylate (a nucleotidyl analogue of acetyl phosphate) with the concomitant formation of PPi.[13]

To verify this conjecture, Berg synthesized acetyl adenylate and verified that acetyl-coA synthetase rapidly and quantitatively converted the compound to ATP in the presence of PPi, or to acetyl-CoA in the presence of added CoA. Hence, the overall reaction could be explained as the result of a single enzyme catalyzing *two* successive steps rather than the three-step process suggested by Lipmann and Lynen.

1. ATP + acetate ↔ acetyl adenylate + PPi
2. acetyl adenylate + CoA ↔ acetyl-CoA + AMP

Satisfied that this observation was not an experimental artifact, Berg was stunned by the realization that the two celebrated biochemists must have erred! Given that the cells from which their enzyme had been obtained were grown on acetate as the carbon source, he inferred that the enzyme preparation they had used contained acetate. "Simply stated," Berg told an interviewer in 1975, "I had discovered a new kind of reaction — that solved a dilemma that had been concerning biochemists before, namely the activation of carboxyl groups. The discovery was essentially that ATP could react with a variety of fatty acids and other

---

* According to Arthur Kornberg, this famous maxim originated with the American biochemist, Ephraim Racker.

compounds containing carboxyl groups in a specific way, thereby generating an activated acyl function which could then undergo a second reaction to produce a variety of compounds."[14]

Soon after these experimental observations were confirmed, Berg presented his findings at a meeting of the Federation of American Societies for Experimental Biology (FASEB) in Atlantic City, New Jersey. For many years, these so-called Federation Meetings, considered the one annual scientific event that no self-respecting biologist could afford to miss, attracted thousands of scientists, postdoctoral fellows and graduate students. Though Atlantic City was better known as a Mecca for high-end gamblers who packed the hotel casinos at night, during Federation Meeting week scientists were the principal occupants of the hotels. The casinos were largely deserted and the famous Atlantic City boardwalk teemed with biologists deep in animated conversation as they scurried from one lecture venue to another. Rumor had it that when the scientists came to Atlantic City, the prostitutes went on vacation!

Prior to the meeting, word of Berg's results had spread in the biochemistry community, promoting a packed lecture theater that included both Lynen and Lipmann. After all was said and done, it turned out that the research teams led by the two renowned biochemists had unwittingly grown yeast in the presence of acetic acid, traces of which inevitably contaminated their relatively crude cell-free extracts. Hence, their reactions were never free of acetate. Though mortified to learn of their error, both scientists were unstintingly graceful. Lynen jokingly told Berg that he was the only person to have ever proven him wrong in an experiment! Lipmann in turn expressed his unabashed embarrassment at having made such a careless mistake!

Berg published a slew of papers on this topic, the weight and impact of which were significantly enhanced by Kornberg's gracious decision not to add his name as a co-author on the published papers, just as Warwick Sakami had when Berg was a graduate student, a gesture that to Berg's knowledge was never repeated in publications from Kornberg's research group. "Had his name been on my papers, there's no question that I would not have benefitted from the discovery to the extent that I did. I have always been profoundly grateful to Arthur for that gesture."[15]

Young readers should recall that in the mid-1950s there was no *Cell* or *Molecular Cell*, the colorful covers of which now grace the shelves of laboratory libraries. The most distinguished general biological periodicals then were *Nature* and the *PNAS* — and perhaps *SCIENCE*. But, as previously mentioned, in the field of biochemistry the *Journal of Biological Chemistry* held sway. In 1956, at the age of 30, Berg's curriculum vitae listed eight publications in peer-reviewed journals. Six were in the *Journal of Biological Chemistry* — and the seventh was in *Nature*.

Years later, at the turn of the century, Berg revisited these epic experiments in an interview with the Hungarian Istvan Hargittai, professor of chemistry at the University of Budapest — and a keen historian of science. "When people ask me, of all the things that I've done in science, which do I have the most pride for, it is *that* discovery, not the one that got the Nobel Prize" he told Hargittai (referring to his subsequent contributions to the discovery of recombinant DNA technology; see later chapters). "Making recombinant DNA was pretty obvious. All one had to do was say: 'I want to do it.' — But a whole new world opened up when the kind of reaction that generates acyl adenylates was discovered. Looking back, the discovery had an enormous impact. Additionally, it was original research, the results weren't predictable, and it transpired when I was really just starting out as a scientist."[16]

Soon after his arrival at Washington University, Kornberg established a joint seminar with the Cori laboratory, a weekly meeting that was magisterially presided over by Carl and Gerty. In due course, Kornberg invited Berg to present a talk at one of these seminars, an occasion usually faced with considerable trepidation by junior speakers. In fact, it was most unusual for postdoctoral fellows to be invited to make presentations at the Cori seminars. "I chose to speak about some of the work in Copenhagen on a second mode of glycogen synthesis," Berg confided.

"People at the seminar were looking at me as if I had lost my mind," he continued, "because the Cori's fame stemmed largely from their discovery that glucose-1-phosphate can generate glycogen by reversal of the glycogen phosphorylase reaction. However, it was known that this reaction is highly disadvantageous, prompting concern among some that it

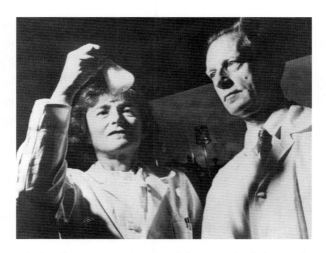

Carl and Gerty Cori (www.nobelprize.org/nobel_prizes)

may not represent the major pathway for glycogen biosynthesis in cells."[17] Notably, several years prior to Berg's sojourn in Copenhagen the Brazilian biochemist Luis Leloir had reported that glycogen synthesis in the liver can be accomplished by transfer of the glucosyl moiety of UDPG (uridyl di-phosphoglucose) to glycogen. Indeed, Berg had specifically elected to present a talk on aspects of glycogen biosynthesis in part because he was aware that Kalckar and several postdoctoral fellows in his Copenhagen laboratory were then examining the enzymatic transfer of glucosyl moieties from UDPG. "To my relief, the Cori's were very decent about the presentation. They were somewhat frigid — but they didn't give me a hard time."[18]

Kornberg had every reason to be pleased with Berg's progress and considered adding him to his growing faculty in the microbiology department as soon as Berg's fellowship terminated. A faculty position was not immediately available, but with help from Washington University, Kornberg secured a three-year appointment for Berg as Scholar in Cancer Research, with the assurance that he would be elevated to the faculty as soon as a position opened. This happy event transpired in 1956, when Berg was appointed Assistant Professor of Microbiology. His earlier disdain for living in St. Louis notwithstanding, it now appeared that Berg might be a citizen of that city for the foreseeable future!

Arthur Kornberg at the blackboard in an obviously jovial mood. (www.google.com/ search?q = arthur+kornberg)

Notes

1. ECF interview with Paul Berg, April 2011.
2. *Ibid.*
3. Jones ME, Lipmann F, Hilz H, Lynen F. (1953) On the Enzymatic Mechanism of Coenzyme A Acetylation With Adenosine Triphosphate and Acetate. *J Am Chem Soc* **75**: 3285–3286.
4. Will H, Hamprecht B. (2011) "Everything Now Seemed So Simple to Me...": Feodor Lynen (1911–1970), a Hero of Biochemistry. *Angewandte Chem* **50**(49): 11580–11584.
5. http://www.bookrags.com/biography/ikipe-lynen-woc.
6. http://history.nih.gov/exhibits/stadtman/bios.htm
7. Kornberg A. (1989) *For the Love of Enzymes: The Odyssey of a Biochemist.* Harvard University Press, p. 82.
8. Jones ME, Lipmann F, Hilz H, Lynen F. (1953) *J Am Chem Soc* **75**: 3285–3286.
9. Judson HF. (1979) *The Eighth Day of Creation: The Makers of the Revolution in Biology.* Simon and Schuster, p. 232.

10. ECF interview with Paul Berg, April 2011.

11. *Ibid.*

12. *Ibid.*

13. PB interview by Sally Hughes, 1999.

14. Berg P. (2008) Moments of Discovery. *Annu.Rev Biochem* **77**: 15–44.

15. *Ibid.*

16. PB interview by Rae Goodell, MIT Oral History Program, May 1975.

17. ECF interview with Paul Berg, April 2011.

18. *Ibid.*

CHAPTER SEVEN

Discovering Transfer RNA

That was my entree to molecular biology

The earliest research to place RNA "somewhere between gene and protein"[1] came in the mid- to late 1930s from the laboratories of Torbjörn
Caspersson at the Karolinska Institute in Stockholm and Jean Brachet at the
University of Brussels; this at a time when studies on gene function had not
moved much beyond the use of specialized histochemistry, a technique that
yielded little more than the crude localization of nucleic acids in cells.
Caspersson documented that most of the RNA in the cytoplasm of mammalian cells was concentrated in minute particles variously referred to as
zymogen granules, small particles, or microsomes.[2] Brachet also identified
such small spherical particles in the cytoplasm of cells and correctly concluded that they contained protein bound up with RNA. He astutely noted
too that cells actively synthesizing proteins were rich in RNA, suggesting
that this nucleic acid was required for protein biosynthesis.[3] But these observations shed no light on the links between gene expression and the synthesis
of a protein. As Francis Crick told historian Horace Judson years later:

> You see. Everything is messy. If you look through the literature on the
> turnover of RNA in cells, you'll find it *exceedingly* complicated and
> difficult and impossible to understand. And one can see why. Because
> you had ribosomal RNA, transfer RNA, and messenger RNA and
> heteronuclear RNA. And, of course, nobody could distinguish them
> at all; they were all muddled up together, except that some was in the
> nucleus and some was in the cytoplasm.[4]

The war years had little use for fundamental biological sciences such as biochemistry, which consequently lost considerable momentum, especially in the UK. But in the summer of 1946, the Society for Experimental Biology at Cambridge University convened a meeting at which both Caspersson and Brachet were present. There was general consensus that whereas DNA is confined to the nucleus, RNA is mainly extranuclear and that a high RNA content is characteristic of cells engaged in vigorous protein synthesis. George Palade's electron microscopic studies of the small RNA-containing cytoplasmic particles (eventually renamed ribosomes by Richard Roberts[5]) confirmed the presence of proteins in addition to RNA, promoting the notion that the genetic code embodied in DNA is communicated directly to ribosomes — and hence to polypeptides — a notion obviously incorrect.

The discovery of messenger RNA (mRNA) by Brenner and Jacob in the early 1960s (and independently by Jim Watson and his colleagues at about the same time) lent key insights into the complexity of protein synthesis. This fundamental breakthrough solved the burning question of how the genetic code is faithfully transcribed to the nucleotide sequence of RNA. But the informational link between mRNA and protein synthesis remained obscure. In 1955, Francis Crick wrote an historic theoretical paper for the RNA Tie Club (an entity comprising about two dozen prominent molecular biologists of the day who circulated newsworthy unpublished manuscripts among themselves) in which he correctly anticipated the existence of a molecular species that served as an indispensable intermediary between mRNA and amino acids in proteins. He wrote:

> The main idea was that it was very difficult to consider how DNA or RNA, in any conceivable form, could provide a direct template for the side-chains of the twenty standard amino acids. What any structure was likely to have was a specific pattern of atomic groups that could form hydrogen bonds. I therefore proposed a theory in which there were twenty adaptors (one for each amino acid), together with twenty special enzymes. Each enzyme would join one particular amino acid to its own special adaptor. This combination would then diffuse to the RNA template. An adaptor molecule could fit in only those places on the nucleic

acid template where it could form the necessary hydrogen bonds to hold it in place. Sitting there, it would have carried its amino acid to just the right place where it was needed.[6]

Crick famously referred to his manuscript (entitled "On Degenerate Templates and the Adaptor Hypothesis: A Note for the RNA Tie Club") as his "most influential unpublished paper!" His notion of an element in protein synthesis (that he called the "adapter"), which specified the order of addition of amino acids to generate new polypeptides, was compelling. But there was no experimental evidence for the existence of "adapters"* nor how they might carry amino acids to "just the right place where it was needed."

In due course, the Crick "adapter" was shown to be a novel type of RNA dubbed transfer RNA (tRNA). The discovery of tRNA is credited to three disparate research groups who independently encountered this missing link not too long after Crick's prophetic manuscript circulated: Paul Zamecnik and his collaborator Mahlon Hoagland at the Massachusetts General Hospital in Boston, Paul Berg in St. Louis and Robert Holley at Cornell University.

By 1953, prior to the discovery of aminoacyl adenylates (see previous chapter), Paul Zamecnik, who had earlier worked with Fritz Lipmann at the Massachusetts General Hospital but was then a professor at Harvard Medical School, had used radioactive amino acids to establish that radioactivity was incorporated into acid-insoluble material — almost certainly protein. These experiments essentially demonstrated that ATP is required for amino acids to be incorporated into proteins, prompting speculation that amino acid activation was imperative.

With this system in hand Zamecnik, together with Hoagland and Elizabeth Keller, established that ATP was required for amino acid

----

* The adapter idea was Crick's prophetic solution to the problem of how amino acids could possibly identify a nucleic acid sequence, i.e. there was no chemical way an amino acid side chain could recognize it's respective coding sequence in an RNA. So he surmised that only a nucleic acid could recognize another nucleic acid by base pairing. He surmised further that amino acids must become attached to a small RNA, and that it is through this adapter that the amino acid "recognizes" its codon.

Paul Zamecnik. (www.emfoley.com/luminaries/index)

incorporation. However, they did not directly demonstrate the existence of aminoacyl adenylates. Several years later Zamecnik and Hoagland showed that a soluble RNA present in a cytoplasmic fraction became radioactively labeled in the presence of ^{14}C-amino acids and that the labeled RNA subsequently transferred the amino acids to proteins. The pair correctly concluded that the RNA in question was an intermediate carrier of amino acids during protein synthesis — and named the entity transfer RNA (tRNA).[7]

In a reflective piece for the journal *Trends in Biochemical Sciences* (TIBS) published in 1996, Zamecnik related:

> It happened that I first announced the discovery of amino acid activation at a meeting of molecular biologists in Detroit in 1955. The palpable indifference with which the audience received the news showed how tightly closed the door between biochemistry and molecular biology compartments was.[8]

Paul Berg independently discovered tRNA by a different route. Having observed that the acetyl-CoA activation system supports ATP-PPi exchange, he began a systematic search for other constituents in extracts of *E. coli* that support this sort of reaction. To his surprise, he observed that the amino acid methionine catalyzed ATP-PPi exchange. Knowing that a specific acceptor (acetyl-CoA synthase) is involved during the synthesis of acetyl-CoA, and having demonstrated that multiple amino acids support ATP-PPi exchange, Berg reasoned that cells may also carry acceptors for the activated amino acids. He set his first graduate student, James Ofengard, to track down this anticipated molecular entity. To Berg's and Ofengard's delight, this turned out to be a small RNA, subsequently shown to be tRNA. "That was my entree to molecular biology," Berg stated.[9]

In mid-November 1957, the US National Academy of Sciences organized a small symposium on amino acid activation. Berg and Zamecnik were two of six investigators invited to present papers, all of which were published in the *PNAS* in February 1957, a few months after the symposium. Berg's and Ofengard's contribution stated:

> We have provided evidence in this communication for an enzymatic reaction linking amino acids to RNA. .... The significance of the present observations on the linking of amino acids to RNA with regard to the mechanism of protein synthesis is not clear. Many of the current ideas concerning protein synthesis visualize this process as involving, first, the activation of the free amino acids to some higher-energy level, followed by the orientation of these activated amino acids in a predetermined sequence on a nucleic acid "template". ....
> .... The most attractive hypothesis at present, although by no means the only one, is that the linking of amino acids to RNA represents the so-called template stage mentioned above. Consistent with this view is the finding by Hoagland *et al.* that the amino acids of an RNA-amino acid compound from rat liver are incorporated into peptide linkage ....[10]

"In essence," Berg noted, "we discovered that amino acids participate in a reaction similar to the one we described for acetate, generating the analogous aminoacyl adenylate. By analogy with the mechanism of acetyl CoA formation I was convinced that there had to be an acceptor for activated methionine. So Ofengand began to search for an entity that would accept activated methionine. To our surprise this also turned out to be a small RNA."[11]

Surmising that the amino acid activation reaction was integral to the assembly of amino acids into polypeptide chains, Berg and Ofengard, joined by postdoctoral fellows Fred Bergmann and Jack Preiss as well as Berg's trusted technician, the late Marianne Dieckmann, revisited the crude system from which they had purified the initial RNA species and identified independent activation systems for each amino acid — all involving different tRNAs.

Recognizing the imperative of demonstrating that the intermediates in the aminoacylation of tRNA were indeed aminoacyl adenylates, Berg spent a summer in the late Gobind Khorana's laboratory in Vancouver, Canada, learning to chemically synthesize a variety of aminoacyl adenylates. With these in hand, each of the purified enzymes was shown to generate ATP in the presence of PPi and to form the respective aminoacyl tRNA. Further experiments established that one or more species of tRNA exists for each amino acid. Moreover, there is only a single amino acyl acceptor site in each tRNA. The stoichiometry of one amino acid per tRNA squared with Francis Crick's supposition that aminoacyl tRNAs serve as "adapters" that match amino acids to their cognate mRNA codons during ribosome-mediated assembly of polypeptide chains.[12]

Berg and his colleagues determined that so-called "amino acid-activating enzymes" were in fact aminoacyl RNA synthetases. Whereas the initial reaction between ATP, amino acid and a specific enzyme results in the formation of an enzyme-bound aminoacyl adenylate, the amino acyl moiety is transferred to the appropriate acceptor tRNA chain. Hence, the series of reactions can be represented as follows:

1. ATP + amino acid → amino acyl-AMP + PPi
2. Amino acyl-AMP + tRNA → amino acyl –RNA + AMP

    ATP + amino acid + RNA → amino acyl-RNA + APM + PPi

Berg *et al.* announced these observations in a seminal paper published in 1959 entitled "The Chemical Nature of the RNA-Amino Acid Compound Formed by Amino Acid-Activating Enzymes."[13] In this paper, the group reported that the amino acids were linked, one to a tRNA chain, either at the 2' or 3' hydroxyl group of the terminal 3' residue of each tRNA. Indeed, Berg was able to generate tRNA preparations with a single functional tRNA specificity by virtue of destruction (with periodate) of all acceptor sites except those occupied by a single amino acid.

"It was really this body of work that established the paradigm that each amino acid is activated by a distinct aminoacyl tRNA synthetase and is then transferred to its cognate acceptor tRNA," Berg stated. He and his colleagues subsequently addressed two other burning questions. One asked what the precise mechanism is by which specific tRNAs are bound to specific amino acids. A second key question asked how amino acyl tRNA's are used in protein synthesis.

In a series of four back-to-back papers published in the *Journal of Biological Chemistry*,[14] Berg and his colleagues announced:

(i) the mechanism of leucyl-tRNA, valyl-tRNA, isoleucyl-tRNA and methionyl-tRNA formation in *E. coli.*

(ii) the preparation of leucyl-valyl-, isoleucyl- and methionyl-RNA synthetases in *E. coli.*

(iii) the isolation of amino acid-acceptor tRNA's from *E. coli.*

Not too much later, Berg, in collaboration with the Swedish biochemist Ulf Lagerkvist, addressed the specificity of each tRNA for a particular amino acid. By the early 1960s, it was shown that each amino acid-acceptor RNA chain contains the identical trinucleotide sequence cytidyl 5'→3', cytidyl 5'→3', adenyl 5'→3' (pCpCpA) at the acceptor end, and a guanosine-5'-phosphate residue at the other end, thereby excluding these elements of the RNA chain as the sole source of the specificity. In considering that specificity might reside in nucleotide sequences adjacent to or near the pCpCpA segment, Berg and Lagerkvist asked whether there was sufficient heterogeneity in these nucleotide sequences to specify 20 or more amino acid side chains, and whether different amino acid-acceptor tRNAs are distinguishable by the nucleotide sequences in this region of the RNA chains.

The pair determined that there are indeed marked differences in the 3'-hydroxy-terminal nucleotide sequences. They identified 11 different sequences, (ranging in length between 4–9 nucleotides) that could account for ~80% of the acceptor tRNA chains. They also established that one of these sequences is present in tRNA chains that accept isoleucine and two others in tRNA chains that accept leucine. But they cautiously concluded: "Although the present study shows that the [tRNA] chains vary in nucleotide sequences next to the CpCpAp end group, we do not know whether this region contributes structural information to distinguish different RNA chains." Two adjacent papers by Lagerkvist and Berg published in the *Journal of Molecular Biology* in 1962 constituted articles v and vi in a series on "The Enzymatic Synthesis of Amino Acyl Derivatives of Ribonucleic Acid."[15,16]

In subsequent studies, Berg, together with postdoctoral fellow Anne Norris, sought an explanation for the observation that in addition to generating an isoleucyl adenylate complex isoleucyl tRNA synthetase can also form a valyl adenylate complex. However, the enzyme only transfers the isoleucyl moiety to the appropriate tRNA, suggesting that tRNA molecules specific for isoleucine react differently with identical protein molecules complexed to either isoleucyl- or valyl-adenylates.

To explore this oddity, Norris and Berg purified isoleucyl tRNA synthetase-bound to the isoleucyl- and valyl-adenylate complexes.[17] The isolated complexes were compared with respect to their ability to transfer the adenylate moiety to PPi with the coincident generation of ATP, and with respect to their ability to transfer the aminoacyl entity to tRNA. Both complexes effectively served as precursors of ATP. But in contrast to the isoleucyl derivative, the synthetase-valyladenylate complex failed to serve as a donor of the valyl group to tRNA. Instead, the complex broke down when reacted with isoleucine-specific tRNA chains. This feature proved to be a general property of multiple aminoacyl tRNA synthetases and provided the first evidence for an error-correcting event during protein synthesis. Since then, error-correcting mechanisms were also shown to exist at the activation step and during the interaction of tRNA's with ribosomes.

In summary, there is no question that Berg's independent discovery of tRNA, and particularly his in-depth exploration of the molecular biology and biochemistry of tRNA function, represent substantial contributions

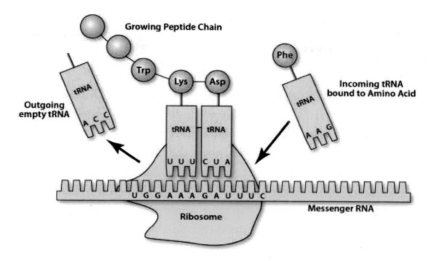

Fig. 1 Schematic representation of the different RNAs required for the biosynthesis of polypeptide chains.

to our understanding of protein synthesis in prokaryotes. All these tRNA studies exclusively deployed the bacterium *E. coli* as an experimental model of tRNA function and specificity.

The third member of the trio involved in the discovery of tRNA, Robert (Bob) Holley, was enjoying a sabbatical leave at Caltech from the New York State Agricultural Station of Cornell University in Ithica NY, where he too was investigating protein synthesis. He came to the conclusion that activated amino acids "must react with something or be transferred to something" and decided to follow the postulated reaction by examining utilization of the energy-rich phosphate bond. In so doing he determined that the reaction was sensitive to ribonuclease, whereas the preceding activation step was not. Before directly identifying the required RNA species, Holley rushed off a report to the *Journal of the American Chemical Society* entitled "An Alanine-dependent, Ribonuclease-inhibited Conversion of AMP to ATP, and its Possible Relationship to Protein Synthesis," a contribution that prompted historian Horace Judson to comment: "........ it reads badly; the evidence was indirect, the argument inconclusive, the work palpably incomplete."[18] "Furthermore", Berg suggested, "it came *after* our own work positing the existence of aminoacyl adenylates and perhaps even after we had synthesized the aminoacyl adenylates chemically and shown that they give rise to ATP in the presence of PPi."[19]

****

About five years after Berg's arrival at Washington University, Arthur Kornberg and the entire microbiology faculty agreed to move to Stanford University to constitute a new biochemistry department (see next chapter). Equally life changing was the news that Paul and Millie were to be blessed by the birth of their son, John Alexander. The couple had hoped for a child for 11 long years, and in 1958 their hopes were finally realized. Six months later they packed up and made the long trek westward to settle in a new home on the Stanford campus, one of the leading universities in the world situated in one of the most climatically pleasant and scenic places in the US. While some of the clutch of publications on tRNA thus far discussed reported work carried out at Washington University, the remainder of the tRNA work — as well as new studies on coupled transcription–translation — were carried out at Stanford.

Notes

1. Judson HF. (1979) *The Eighth Day of Creation: The Makers of the Revolution in Biology*. Simon and Schuster, p. 234.
2. http://en.wikipedia.org/wiki/Torbjörn_Caspersson;
3. http://en.wikipedia.org/wiki/Jean_Brachet
4. Judson HF. (1979) *The Eighth Day of Creation: The Makers of the Revolution in Biology*. Simon and Schuster, p. 236.
5. http://www.newworldencyclopedia.org/entry/Ribosome; Roberts RB. (1958) Introduction. In: Roberts RB, *Microsomal Particles and Protein Synthesis*. New York, Pergamon Press.
6. Crick F. (1988) *What Mad Pursuit. A Personal View of Scientific Discovery*. Basic Books, p. 96.
7. *JBC Classics*. (2005) 280: p. e37-e39.
8. Zamecnik P. (1996) *TIBS* **21**: 77–80.
9. ECF interview with Paul Berg, February 2012.
10. Berg P, Ofengand EJ. (1958) *PNAS* **44**: 78.
11. ECF interview with Paul Berg, February 2012.

12. Berg P. (2008) Moments of Discovery. *Annu Rev Biochem* **77**: 15–44.
13. Preiss J, Berg P, Ofengand EJ, *et al.* (1959) *PNAS* **45**: 319–328.
14. Berg P, Bergmann FH, Ofengand EJ, Dieckmann M. (1961) *J Biol Chem* **236**: 1726–1734; Bergmann FH, Berg P, Dieckmann M. (1961) *J Biol Chem* **236**: 1735–1740; Ofengand EJ, Dieckmann M, Berg P. (1961) *J Biol Chem* **236**: 1741–1747; Preiss J, Dieckmann M, Berg, P. (1961) *J Biol Chem* **236**: 1748–1757.
15. Lagerkvist U, Berg P. (1962) *J Mol Biol* **5**: 139–158.
16. Berg P, Lagerkvist U, Dieckmann M. (1962) *J Mol Biol* **5**: 159–171.
17. Norris AT, Berg P. (1964) *Fed Proc 23*, No. 2.
18. Judson HF. (1979) *The Eighth Day of Creation: The Makers of the Revolution in Biology*. Simon and Schuster, p. 326.
19. ECF interview with Paul Berg, February 2012.

CHAPTER EIGHT

Stanford University — and its Refurbished Department of Biochemistry

Kornberg sought to maintain a relatively small faculty

The discovery of gold in California in 1848 was followed by a flood of "gold-diggers," who populated the state from 1849 (hence the term "forty-niners") into the 1850s. This incursion included two extraordinarily forward-looking and dynamic personalities: Elias Samuel Cooper and Leland Stanford, both destined to shape the future of the Stanford University School of Medicine.

Leland Stanford grew up and studied law in New York before moving west after the gold rush. He was one of four major California businessmen known popularly as the "Big Four" (or among themselves as the "Associates"), crucial investors in the formidable challenge of building the Central Pacific Transcontinental Railroad linking the western states with the rest of the country. In so doing the foursome amassed financial fortunes.*[1] In 1876,

---

*In later years, his use of cheap Chinese labor for this grueling task led to his identification by Stanford students as a "robber baron." Indeed, when the Stanford University football team relinquished the politically incorrect nickname "the Stanford Indians" in 1972, Stanford students offered the moniker "the Stanford Robber Barons" as a replacement. This snide suggestion was ignored of course, leading to the team's present designation as the Stanford Cardinal, after the color of the uniforms worn by all Stanford athletes.

Stanford (variously a leader in the Republican Party, Governor of California and later a US senator) purchased 650 acres of Rancho San Francisquito for a country home and began the development of his famous Palo Alto Stock Farm, where he bred thoroughbred horses.[†] He later purchased adjoining properties totaling more than 8,000 acres and the small town that was beginning to emerge nearby took the name Palo Alto (tall tree) after a giant California redwood on the bank of San Francisquito Creek.[‡]

Stanford and his wife, Jane, had one child, Leland Stanford Jr., whose life was tragically cut short by a bout of typhoid while the family was traveling in Italy. The lad was just 15 years old. Within weeks of his death, the bereaved Stanfords announced their intention to establish a new university in California. "The children of California shall be our children," was their mantra then. Indeed, the (rarely used) proper name of this august educational establishment is Leland Stanford *Junior* University. In 1885, less than two years after Leland Jr.'s death, the Stanfords legalized a grant of endowment. Construction of the university began with the traditional laying of a cornerstone on 14 May 1887, the day the young man would have celebrated his 19th birthday.[2]

In September 1891, Leland Stanford announced that he intended to establish a medical school in the University sometime in the future. A year later, a story appeared in the *San Francisco Examiner* to the effect that the University of California and Stanford University "are both striving by every possible means to secure Cooper Medical College, a medical school founded by Elias Samuel Cooper in San Francisco with

---

[†] In 1872, Stanford commissioned Eadweard Muybridge to use newly invented photographic technology to definitively determine whether a galloping horse has all four feet off the ground simultaneously. It does! This project, which illustrated motion through a series of still images, was a prelude to the technology that initiated motion films. [http://en.wikipedia.org/wiki/Leland_Stanford].

[‡] The El Palo Alto Redwood, now a California Historical Landmark, is well over 1,000 years old and is recognized for its historical significance as a campsite for the expedition led by the Spanish explorer Gaspar de Portolà in 1769. [http://en.wikipedia.org/wiki/Portolà_expedition] Surveyors plotting out El Camino Real, the main street in Palo Alto, also used it as a sighting tree. The tree is a centerpiece of the Stanford University seal and has been adopted as the unofficial university mascot. [http://www.cityofpaloalto.org/default.asp]

Leland Stanford with his wife, Jane Lathrop Stanford, and son, Leland Stanford Jr. (www.digitalhistoryproject.com/2012/09/jane.lathrop)

a rich history — but without university affiliation."[3] This conjecture prompted Stanford University president David Starr Jordan to write disappointingly:

> There is nothing as yet in the discussion of the union of Cooper Medical College [with Stanford University]. It seems to have started in the City without any provocation on our part. ⋯⋯⋯ I do not think Mr. Stanford wishes to extend the University in the direction of medicine for the present.[4]

Further speculation regarding this exciting development fizzled following Stanford's death in June 1893, an event that plunged the institution into financial crisis because it had not been incorporated separately from Stanford's other properties, and its assets and income were tied up in prolonged probate proceedings. Compounding this problem, the federal government filed suit against Stanford's estate for his presumed share of construction loans made to the Central Pacific Railroad. These financial concerns notwithstanding, Stanford's widow Jane, a remarkable woman by any yardstick, stepped up to the plate. She categorically refused to close the university and undauntedly funded faculty salaries and university operations from an allocation made to her by the probate judge — and not infrequently from her own pocket. The estate was released from probate in 1898 and following her sale of the railroad holdings the following year, Mrs. Stanford turned over $11 million to the university trustees. What president Starr Jordan identified as "six pretty long years" had finally come to a close. During that time, Jordan proclaimed, "the future of a university hung by a single thread — the love of a good woman."[§5]

The shared interest in higher education between president Jordan (an eminent natural scientist) and Dr. Eli Cooper Lane (president of Cooper College and nephew of its founder, Elias Samuel Cooper) fueled the possibility of interactions between the two institutions. This expectation was heightened when it seemed certain that all future medical schools in the United States would be obligated to be integral parts of universities.[6]

Eli Cooper Lane was initially opposed to the notion of his beloved San Francisco-based medical school being incorporated into a larger university. But by the turn of the century he began to accept the inevitable reality that the future of medical education in the United States was to be under the control of universities. He also recognized that the cost of operating his own college would become prohibitive once it became necessary

---

[§] Though most history books attribute Mrs. Stanford's death at 76 to heart failure, a closer look at the documents and drama surrounding her demise reveals a quite different picture. A Stanford physician, the late Robert W.P. Cutler, convincingly answered the question of how Mrs. Stanford met her end in a book published in 2003: *The Mysterious Death of Jane Stanford* (Stanford University Press). Cutler claims she was murdered by strychnine poisoning whilst on vacation in Hawaii. Cutler also espoused the view that university president Jordan went to extreme lengths to cover up this event. [http://alumni.stanford.edu/get/page/magazine/article/?article_id=36459]

to appoint salaried professors.[7] In August 1906, the Stanford trustees appointed a committee to discuss consolidation with the leadership of Cooper Medical College. The minutes of the Stanford Board of Trustees for August 1, 1906 include the following relevant entry:

> In its report the committee recommended that all assets of Cooper Medical College, including its hospital, equipment and cash funds be conveyed to Stanford University, that an equitable agreement be arrived in naming the medical school and that "in the event that the Board of Trustees of the Leland Stanford Junior University should fail for any cause to maintain a Medical Department for the purposes expressed, the property shall revert to the State of California for the maintenance of medical education."[8]

The Stanford trustees formally approved the merger in October 1907. For the 51 years between 1908 and 1959, instruction at Stanford Medical School consisted of two years of basic science teaching on the Stanford campus, followed by two years of clinical training at the San Francisco facilities Stanford inherited from Cooper Medical College. In 1953, this unwieldy and awkward academic structure was slated for reorganization when the Stanford trustees announced that the university would establish a comprehensive medical school on the Stanford campus. Edward Durell Stone was selected as the architect for Stanford's $21.5 million, 440-bed Palo Alto-Stanford Medical Center. Comprised of three hospital and four medical school buildings interconnected by numerous arcades and open walkways, the strikingly beautiful 56-acre site was designed along lines similar to those of Stanford's imposing main quadrangle. A three-story height was maintained throughout, and the concrete walls and columns of the center were patterned as a grillwork to simulate the sandstone-block surfaces of the main university Quad.[9]

Soon after Robert (Bob) H. Alway, professor and chairman of the Stanford Department of Pediatrics accepted the deanship of the medical school, university president Wallace Sterling announced that all existing medical school department heads were expected to resign by August 31, 1958, a strategic move undoubtedly hatched by Alway's perception that there had been too much inbreeding in developing leadership positions in the school.[10] Always lost little time in searching out new talent and quickly recruited an impressive cadre of replacements for vacated and vacant departmental chairs. Almost overnight, Stanford Medical School was catapulted to national prominence. Alway also created a new Department of Genetics that he populated with a few full professors. But he delayed announcing a chairperson for this new entity.

In April 1957, Henry Kaplan, an authority on radiotherapy and chairman of the Stanford Department of Radiology informed Arthur Kornberg of the university's keen interest in him for the chairmanship of its new biochemistry department. Two months later, Kornberg and his wife Sylvy were invited to visit. In his autobiography *For the Love of Enzymes — The Odyssey of a Biochemist*, Kornberg related that he and his wife "tried to look beyond the stunning reception — beyond the flattering fact, for example, that when they stopped for gas at Rickey's Motel on El Camino Real (a main street in Palo Alto) where they were housed in an elegant suite, the attendant is said to have stated: 'By the way, Dr. Kornberg, I enjoyed your seminar very much!'"[11]

On returning to St. Louis, Kornberg communicated the Stanford prospects to his faculty colleagues in the department of microbiology. Years later he wrote:

> I had to admit that the case was compelling. The Stanford Medical School would be moving in two years from San Francisco to the campus in Palo Alto. In the handsome new buildings designed by Edward Durell Stone ···················· the biochemistry department was assigned a spacious floor that I could design to my own specifications and the unpopulated department would be staffed with colleagues of my choosing. Now a legitimate biochemist, I could teach and practice my subject without disguise and appoint the physical and organic biochemists whom I very much missed. No longer would I be the muted, junior member of the tradition-bound Executive Committee of Washington University Medical School. Rather, I would have a large share in policy and faculty selection in the renaissance of science and medicine at Stanford promised by J. E. Wallace Sterling, the President, and Frederick E. Terman, the Provost. Having escaped to California from St. Louis for several summers, Sylvy and I were now also keen to live permanently in that

most agreeable climate and geographical setting, one that would also be attractive to our children, future students, and colleagues. Instead of remaining in the 'Gateway to the West,' we might now be housed in the manor itself.[12]

"The two lame-duck years in St. Louis were awkward," Kornberg wrote. "There were some resentments among the Washington University faculty, despite our having brightened and equipped the place, largely with external grants — and having provided a clean slate for appointment of a new faculty. In June 1959, 22 people and five children set out for the promised land."[13] The biochemistry faculty was subsequently further enriched by the recruitment of George Stark and Lubert Stryer, followed by Ronald (Ron) Davis and Douglas (Doug) Brutlag and subsequently by others.

Kornberg let it be known to Dean Alway that he intended to move the entire faculty of the Department of Microbiology at Washington University with him — and that this issue was non-negotiable! Not surprisingly, a decision to move to California and to occupy an entire floor of custom-designed space in a brand new building on the campus of one of America's great universities required little reflection by his faculty colleagues in St. Louis. Reciprocally Alway was delighted to add the likes of Paul Berg, Mel Cohen, David Hogness, Dale Kaiser and Bob Lehman to his faculty. Additionally, Robert (Buzz) Baldwin, recognized for his contributions to the physical chemistry of proteins, agreed to join the Stanford department from Wisconsin. And upon hearing of these startling developments, the accomplished young biologist Charles Yanofsky, then at Western Reserve University, accepted an offer from the Stanford Department of Biological Sciences. Kornberg also actively participated in the recruitment of new faculty for the Stanford Department of Chemistry, which, "with the selection of William Johnson as chairman and then Carl Djerassi, Paul Flory, Henry Taube, Harden McConnell, and Eugene van Tamelen, also vaulted to national prominence."[14]

The move was not without its share of hiccups. To Kornberg's dismay, his introduction to his new department mirrored the experience received when he moved to Washington University — the physical construction of the new department was not complete. Kornberg, who was renowned for his intolerance of roadblocks and hindrances, used some of his "credit" as a recently recruited chairman to light a few fires. In no time workmen from all over the medical center were transferred to the third floor of the new medical school.

Once renovations were completed, truckloads of boxes containing precious scientific material of one sort or another had to be carefully unpacked and put away. Impressively, the departing Washington University faculty had packed up the laboratory with considerable foresight, lightening the potential for unnecessary confusion at the Stanford end. Carpenters at Stanford were instructed precisely how and where to build drawers and cabinets. "We had every barrel, every box marked and ticketed almost exactly with the drawer or shelf its contents were supposed to go into," Berg related. "It was a really amazing move. Everything was planned down to a T."[15] About six weeks after occupying the Stanford department's spanking new space, research in the faculty laboratories was underway.

Despite pressures and temptations to the contrary, Kornberg maintained a small faculty group, about half the number of that in comparable biochemistry departments around the country. Rather than trying to blanket numerous disparate subject areas, he concentrated on essentially one theme: nucleic acids and the proteins with which they interact. "Yet, I felt our scope was broad because we embraced a spectrum of disciplines, from physical chemistry to genetics," he wrote.[16]

The rapid rise to prominence of a new biomedical research entity on the West Coast generated considerable interest outside the corridors of Stanford. Wary of the future of an unknown medical school, Joshua (Josh) Lederberg, who had recently won a Nobel prize and was being heavily recruited by a number of academic institutions, initially declined an offer to assume the chairmanship of the new Department of Genetics. But when he learned of Kornberg's plan to move lock stock and barrel to Palo Alto, he was persuaded to rethink his decision. "We're all coming from St. Louis, and between your genetics and our biochemistry, we can really be a powerhouse," Kornberg enthusiastically informed Lederberg. Stanford University was of course thrilled to have landed both Arthur Kornberg, who won a Nobel prize the very year of the move, and Josh Lederberg, who had gained that singular distinction the year before.

There being no suitable space to immediately house a new genetics department, Lederberg and a fledgling group of faculty initially camped out in space generously lent by the Department of Biochemistry. But the two departments operated differently. "People in the department of bio-chemistry were extremely interactive," Berg related. "Josh's philosophy on the other hand was essentially 'you're on your own'". He was frequently traveling around the world doing his thing, both scientific and political, and his young faculty had to fend for themselves. So the anticipated coalescence between the departments of genetics and biochemistry never really materialized. It didn't transpire with respect to teaching — and it didn't happen with research. As discussed more fully later in this book, interspersed as the biochemistry and genetics labs were, some research and intellectual interactions were inevitable. We knew them and we knew what they were doing,"[17] Berg stated. Interactions progressed when Stanley Cohen arrived with a joint appointment in the Department of Genetics and the Department of Medicine's division of clinical pharmacology. His work on microbial drug resistant plasmids promoted interactions with several individuals in the Department of Biochemistry, specifically Mort Mandel, a visitor in Dale Kaiser's lab and with Berg's graduate student, Janet Mertz.

In his extended interview by Sally Hughes in 2000, Berg conceded that Kornberg's calculated philosophy of self-containment escalated to a reputation for aloofness among some members of the Stanford basic science community — and beyond. "Snootiness" and "arrogance" were other descriptors that circulated. "The reality is that the department really didn't interact with many outsiders," Berg candidly stated.[18] Notably, Kornberg did not favor joint appointments except when Eric Shooter was recruited to the Genetics Department faculty and was invited to a joint appointment in Biochemistry.

When Berg was chairman of the biochemistry department (1969–1974), he sat on several committees that were striving to create a cancer center. The notion was that there would be a new building that would house cancer biologists drawn from the entire medical school. Berg felt compelled not to support that model because his faculty "were unwilling to be located anywhere other than in the department proper!"[19]

All this being said, there is little question about the national and international visibility of the Stanford Department of Biochemistry for many years. Both Kornberg and Berg brought Nobel prizes to the institution, and every member of the faculty group that arrived from St. Louis was in due course elected to membership in the US National Academy of Sciences. Additionally, despite the reluctance to allow faculty with primary appointments in other departments to "worship in the temple," many at the medical school (including this author) enjoyed profitable interactions with the biochemistry faculty, postdoctoral fellows and graduate students.

Kornberg's notable scientific accomplishments were complemented by a reputation (deserved or not) for his unyielding views on the fundamental importance of basic science when approaching complex biological phenomena. He was of the entrenched view that understanding the basic enzymological details and mechanisms of simple biological systems, notably bacteria (especially *Escherichia coli*), would shed light on more complex systems. His passion for *E. coli* as an experimental model, one that he mined most productively, was indeed intense. "If one was interested in immunity for example, one would surely select a model system in which immunity exists," Berg argued sardonically. "But Arthur would never concede such logic. If confronted with such a hypothetical he would almost certainly argue that mechanistic principles learned from *E. coli* might be key to understanding immunity in higher organisms — even though bacteria don't have an immune system! He had the same attitude about working with tumor viruses and mammalian cells in order to understand cancer. He talked about 'fashions in research' and lamented the contemporary vogue of discarding simple prokaryotic systems that had been so informative over the past 30 to 40 years."[20]

Kornberg remained chairman of the biochemistry department for the first 16 years after the move to Stanford, following which he felt the department, then established, would benefit from new leadership.

Importantly, he was aware that Berg was eager for that role — and he championed his appointment as chairman, an appointment that began in 1969.

Berg enjoyed a close relationship with Kornberg, a relationship that began when he was a postdoctoral fellow in Kornberg's laboratory in St. Louis. "Arthur took a liking to me from the first time I walked into his lab in St. Louis," Berg related. "Maybe it was partly because I was the only one of five postdocs who decided to join his lab at Washington University! In any event, when Millie and I arrived in St. Louis the Kornberg's made a huge fuss over us. They lent us money and helped us find a place to live. And though I was a mere post-doctoral fellow he and I used to often talk broadly about science."[21] Berg's warm personal relationship with Kornberg matured when in 1959, the Kornberg and Berg families both relocated from St. Louis to Palo Alto, California. Indeed, as they grew up, Kornberg's three sons, Roger, Tom and Ken, viewed Paul and Millie Berg as family.

Like many biochemists of his generation, Kornberg failed to fully appreciate the importance of genetics and the role it was playing in the latter part of the 20th century as biology began to focus on more complex systems. Indeed, few biochemists of the day realized that the philosophic tensions that then existed between the disciplines of genetics and biochemistry essentially emulated those between biologists and chemists in earlier years.

Berg related an interesting (and amusing) example of these reservations when he and Kornberg were interviewing Cynthia (Cyndy) Kenyon for a faculty position in biochemistry. A product of the MIT graduate school program and at that time a postdoctoral fellow completing an impressive body of work on the developmental biology of the nematode *C. elegans* at the Laboratory of Molecular Biology in Cambridge, UK, Kenyon was widely considered a rising star. She had been invited to Stanford to present a research seminar, a not infrequent prelude to consideration for a faculty position. As part of her seminar she described an elegant genetic approach to obtaining mutants of the worm that might yield important developmental insights. In the course of a private conversation, Kornberg asked Kenyon why the she was interested in joining a biochemistry department if she was in fact a developmental biologist exploiting genetics. Kenyon politely explained that the genetic approach she was using was merely a prelude to understanding developmental mechanisms at the molecular level. When Kornberg then provocatively asked why the biochemistry department at Stanford should be interested in her, Kenyon blithely responded: "I didn't ask to come here; you invited me!" History does not reveal Kornberg's response to that reproach. In the end, Kenyon went on to a stunningly successful career at the nearby University of California at San Francisco (UCSF).[22]

Kornberg also acquired a reputation for determined single-mindedness and frank stubbornness about some issues. Hence, when Berg informed Kornberg of his decision to spend a year in Dulbecco's laboratory (see prologue) working with, of all things, animal cells and viruses, he received a disappointingly frigid reception. "Everyone I had spoken to thought that switching my research focus was a great idea," Berg commented. "But the plan didn't meet with Arthur's approval. He more or less ranted: 'You have such a gift for enzymology! You're wasting your talent! You're destroying your career! The only true path to knowledge is *E. coli*' — and so on. "I won't say we ever came to blows," Berg jokingly continued, "but there were times when I was absolutely furious with Arthur because he could be so

Arthur Kornberg and Paul Berg at the opening of the Arthur Kornberg Medical Research Building at the University of Rochester, Sept. 1999. (Courtesy of Paul Berg.)

critical and narrow-minded. He predicted total failure with my proposed research endeavors and he referred to me as a Pied Piper leading others in the department astray and luring them away from important basic research into a complex field beset with uncharacterized and complicated model systems."[23]

Nor did Berg view his reorientation from bacteria to mammalian cells as an especially drastic transition. "The bacterial field, particularly in the area of gene expression, was rapidly clarifying," he stated. "The predominant paradigm for understanding gene function in lower organisms was essentially the Jacob–Monod story.[*] But essentially nothing was known about the regulation of gene expression in mammalian cells. So it seemed perfectly natural to me to ask whether knowledge gained from bacterial systems might illuminate our understanding of gene regulation in eukaryotes," Berg related. "I was ambitious, I considered myself reasonably bright and I had already made a mark in biochemistry circles. Nor was this exploratory attitude novel among the biochemistry faculty. Notably, when he abandoned phage λ and adopted myxobacteria as a model organism Dale Kaiser cultivated a new experimental system in developmental biology. And David Hogness, who had worked extensively on bacteriophage, switched to Drosophila genetics. Both became leaders in new fields that they literally created."[24]

In spite of this occasional petulance, there was another side to the Kornberg coin. Aside from his heroic efforts on behalf of the Stanford Genetics and Chemistry Departments in the late 1960s, Kornberg went out of his way to promote the recruitment of David Korn, a trained pathologist who was then working on phage λ at the NIH. When he assumed the chairmanship of the Stanford Department of Pathology, Korn switched his own research focus to the biochemistry of DNA polymerases from higher organisms. He lost no time in recruiting a strongly research-oriented group of faculty, a number of whom went on to

---

[*] In 1961, François Jacob and Jacques Monod, two French biologists working at the Pasteur Institute in Paris, France correctly postulated the essential events involved in translating the genetic code to proteins, and terms such as "Jacob–Monod hypothesis," or simply "a la Jacob and Monod," refer to the general mechanism of protein synthesis in bacterial cells. [http://www.bookrags.com/research/jacob-monod-hypothesis-wob]

distinguished careers as molecular pathologists. While joint appointments in the Department of Biochemistry were never on the table, the pathology faculty from Korn down frequented the biochemistry halls with gay abandon, and interactions with the "real" biochemists were in the main relaxed, friendly and typically of significant value. In a nutshell, from the time of his chairmanship in the Department of Microbiology at Washington University, Kornberg sought to maintain a relatively small faculty that functioned as an intellectual family with common research goals and interests. Outsiders were welcomed as visitors — but the Stanford department's faculty did not grow further until Berg assumed the chairmanship in 1969 and Ronald Davis and Douglas Brutlag were appointed as assistant professors. Lamenting the increasing size of the department, Kornberg related:

> ········ some precious things were lost in this otherwise healthy growth. One, which I fondly recall, was the monthly evening research seminar, with everyone crowded into the living room of my house. After several years we had to move the meeting to the larger library space in the department and then to the gazebo in the chemistry department. An essential ingredient of intimacy was lost and the monthly seminars quickly vanished.[25]

During Berg's tenure, the monthly departmental research meetings were supplanted by biennial and subsequently annual research conferences at the Asilomar Conference Center in Pacific Grove, California, a picturesque facility that served an important role in the recombinant DNA controversy (see later chapters).

Importantly, however, for all the criticisms leveled at "King Arthur" over the years, both by those in his department and by less informed individuals outside the department, and even outside Stanford, he was held in enormous esteem and affection by all. In particular, while he and Berg occasionally differed on matters of policy and even on directions in science, Berg has always held enormous admiration, "bordering on deeply felt love" for his mentor, who occupied a special place in both his professional and personal lives.

Notes

1. http://elane.stanford.edu/Wilson/index.html
2. *Ibid.*
3. http://quake06.stanford.edu/centennial/gallery/people/stanfords/index.html
4. http://elane.stanford.edu/Wilson/index.html
5. *Ibid.*
6. *Ibid.*
7. *Ibid.*
8. *Ibid.*
9. *Ibid.*
10. *Ibid.*
11. Kornberg A. (1989) *For the Love of Enzymes: The Odyssey of a Biochemist.* Harvard University Press, p 180.
12. Kornberg A. (1989) *For the Love of Enzymes: The Odyssey of a Biochemist.* Harvard University Press, p 181.
13. *Ibid.*
14. Kornberg A. (1989) *For the Love of Enzymes: The Odyssey of a Biochemist.* Harvard University Press, p 183.
15. PB Interview by Sally Smith Hughes, Program in the History of Biosciences and Biotechnology, University of California (1997).
16. *Ibid.*
17. ECF interview with Paul Berg, April 2011.
18. *Ibid.*
19. *Ibid.*
20. PB Interview by Sally Smith Hughes, Program in the History of Biosciences and Biotechnology, University of California (1997).
21. *Ibid.*
22. PB Interview by Sally Hughes, Regional Oral History Office of the Bancroft Library at the University of California, Berkeley, 2000.
23. *Ibid.*
24. ECF interview with Paul Berg, April 2011.
25 *Ibid.*

CHAPTER NINE

Transcription and Translation: New Directions

The potential of unraveling the genetic code

In 1959, Berg's scientific successes were acknowledged through the prestigious Eli Lilly Award in Biochemistry, the California Scientist of the Year award — and by coveted election to membership in the US National Academy of Sciences. Election to the American Academy of Arts and Sciences was just a year away. Paul Berg was now a "hot property" in scientific circles, and in succeeding years he was inundated with job offers on a regular basis. But he honored an unwavering commitment to Stanford University — and everything the institution stood for since

Paul Berg (circa late '70s or early '80s). (Courtesy of Paul Berg.)

arriving there. Accordingly, neither he nor his wife Millie experienced any desire to leave the California Bay Area. The closest he ever came to even considering relocating was in late 1977, when he paused briefly to reflect on an offer to assume the presidency at the California Institute of Technology (Caltech) in Pasadena, California. He was fully aware of the exceptional history and reputation of this famed Southern Californian bastion of academic excellence, and for a while he was tempted – though never sorely so. On December 27, 1977 he composed a polite and thoughtful response to Stanton Avery, Chairman of the Caltech Board of Trustees:

> Christmas 1977 has not been a restful holiday; it has been a time of turmoil and decision. The question you and your colleagues brought me ·············· was unexpected, flattering and difficult. Though I leaned heavily toward saying no prior to your Stanford visit, the sincerity and depth of your interest persuaded me to reconsider and start anew.
>
> Searching discussions with my wife, a few close friends and colleagues, but most of all with myself, have occupied the last twelve days. Responsibilities, loyalties, opportunities, ambitions and failings were closely scrutinized and reexamined. I wrestled with thoughts of what stimulates and bores me, what I'm patient and impatient with, and the day-to-day activities of professor-scientists and college presidents. As expected there were no easy answers but there was also no avoiding a decision.
>
> I have decided to ask you not to consider me further as a candidate for the Presidency of the Institute. I am confident that is the proper decision for me.[1]

****

The Stanford Department of Biochemistry welcomed its first graduate students in September 1959, shortly after the group from Washington University had settled into its new home. Among them, Michael (Mike) Chamberlin and William (Bill) Wood, both Harvard undergraduates who ultimately progressed to outstanding scientific careers, elected to join Berg's laboratory as graduate students.

As mentioned in Chapter 7, the presence in living cells of an RNA species called messenger RNA (mRNA) had been announced by Sydney Brenner and the late Francois Jacob of the Pasteur Institute in Paris. It was reasonably presumed that mRNAs were transcripts of genes in DNA. Hence, it naturally followed that a specific enzyme that catalyzes transcription must exist. In the early 1960s, multiple laboratories began offering reports of an enzyme called RNA polymerase in plant, bacterial and animal sources, which in the presence of DNA and the four naturally occurring ribonucleotide triphosphates supported the synthesis of an RNA moiety.

Chamberlin decided to search for an RNA polymerase in extracts of *E. coli* — and to purify this putative enzyme. He anticipated that with Arthur Kornberg working on DNA polymerases nearby, he stood to obtain valuable advice and guidance. Chamberlin soon isolated a nearly homogeneous enzyme that he used to study the nature of the transcription reaction.[2] Bill Wood, on the other hand, elected to pursue the question of how aminoacyl-tRNAs function during protein synthesis. He set out to develop an efficient cell-free system for measuring protein synthesis and toward that goal he generated a fairly "clean" soluble supernatant fraction enriched for ribosomes, with which he was able to demonstrate significant but not especially robust levels of amino acid incorporation into protein. Chamberlin and Wood subsequently teamed up to consolidate this observation by asking whether mRNA synthesized using T4 phage DNA as the transcription template would stimulate protein synthesis. It did so dramatically, prompting Wood to explore coupled *in vitro* transcription–translation in greater detail.[3]

It did not take long for Berg, Chamberlin and Wood to appreciate the potential of this coupled cell-free transcription–translation system for deciphering the genetic code. If, for example, instead of using DNA as a template for RNA synthesis one used a defined homopolymer [e.g. poly(dC)], one would anticipate the synthesis of poly(rG), which in turn would be expected to be translated to the amino acid glycine. "In essence we were tactically well positioned to comprehend essential elements of transcription and translation in prokaryotes, the two major aspects of gene expression — and perhaps even begin to comprehend the genetic code that specifies each amino acid," Berg stated.[4]

Providentially, Berg had been invited to give a lecture at the 1961 International Congress of Biochemistry in Moscow and eagerly anticipated presenting these experimental observations, especially announcing his expectations that the studies in his laboratory had the potential of unraveling the genetic code. To his disappointment — but unstinting admiration — he was scooped by what he described as an "electrifying announcement at the meeting in Russia by Marshall Nirenberg from the NIH, that synthetic RNA polymers directed the formation of amino acids in a specific manner."[5]

To add insult to injury, the scientific session in which Berg was scheduled to speak was precipitously cancelled in deference to celebrating the return of the Russian cosmonaut Vladimir Titov from space! The Russian government blithely co-opted various meeting rooms reserved for the international meeting for press conferences. Scientific presentations were suspended — and never reconvened.

Nor was visiting Russia in 1961 particularly gratifying from a logistical perspective. The Cold War between the Soviet Union and the US was at its height, and American scientists were accorded little courtesy in Russia. When, after a frustrating struggle in the face of an aggravating informational vacuum Berg eventually located the hotel in which he was booked to stay, the authorities held his passport for the duration of his visit (not an unusual event when visiting foreign countries). But retrieving it on his departure from the massive pile of foreign passports carelessly tossed into a drawer in no particular order turned out to be a veritable nightmare.

The dark side of Russian governmental oversight in the early 1960s surfaced again when the eminent French scientist Marianne Grunberg-Monago, a friend and scientific colleague of Berg's, invited him and several other American scientists to dine at her home.[6] Grunberg-Monago was born in Russia. She left the country with her parents at the age of nine, but still had family there. She discovered an enzyme called polynucleotide phosphorylase that proved to be critical for Nirenberg's experiments and in later years served as president of the French Academy of Sciences.

By tacit agreement of her guests, no politically related conversation transpired on the occasion of dinner at her home. A few days later, Grunberg-Monago scheduled a visit to her family's country dacha in a place that Berg described as "an 18th century village." No sooner had the party arrived at their appointed destination than Grunberg-Monago suggested a walk in the countryside. "We walked a considerable distance before Marianne felt secure enough to unburden herself about the political situation in the country under Nikita Khrushchev's charge," Berg related. "That was an illuminating diversion from the science!"[7]

A final adventure with Russian paranoia and intransigence transpired when Berg was milling around the airport (which he described as an absolute zoo) waiting for his plane to depart — an event not to taken for granted! At one point he decided to contact Millie (who was then revisiting Copenhagen) to inform her of his anticipated departure. Access to a telephone proved to be another impenetrable logistical challenge. So Berg composed a brief telegram instead. "Dear Millie," the telegram read, "Can't tell you how much I miss you! Returning on SAS flight —. Love, Paul." When Berg delivered his intended telegram to the official charged with dispatching them, the individual read it, looked at him with obvious suspicion, and asked: "Why can't you tell her?" It took all that Berg could do to keep a straight face! When the SAS flight finally lifted off Russian soil, Berg and his fellow passengers broke out into spontaneous applause![8]

Notes

1. Letter to Stanton Avery, December 27, 1977, Reproduced with permission from the Department of Special Collections, Stanford University Libraries.
2. Chamberlin M, Berg P. (1962) *PNAS* **48**: 81–94.
3. Wood WB, Berg P. (1962) *PNAS* **48**: 94–104.
4. ECF interview with Paul Berg, September 2011.
5. *Ibid.*
6. http://en.wikipedia.org/wiki/Marianne_Grunberg-Manago
7. ECF interview with Paul Berg, April 2011.
8. *Ibid.*

PART III

PART III

CHAPTER TEN

Making Recombinant DNA — The First Faltering Steps

One morning Paul and I got together and he suggested trying to put new genes into SV40

The events related at the end of the prologue left Berg disappointed. His scheme of capitalizing on the (theoretical) potential of naturally generated SV40 pseudovirions as a source of hybrid SV40-bacterial DNA was not practicable. Back to square one as it were, after further pondering he devised an alternative scheme — one that brought to bear his extensive expertise in nucleic acid biochemistry by exploiting some of the rich collection of enzymes tucked away in the biochemistry department freezers.* Many of these enzymes were known to utilize DNA as a substrate in various distinctive biochemical reactions. Instead of relying on Nature to do the work of joining bits of bacterial and viral DNA, Berg's new plan addressed the premeditated covalent joining of fragments of DNA from SV40 virus and the bacterium *E. coli* in the test tube. Hence the moniker *recombinant DNA technology.*

This expression requires some nomenclatural clarification. Recombination in *living* cells (typically referred to as *genetic recombination*) involves

---

* Many of these precious enzymes were the result of the labors of graduate students who were required to generate these reagents as part of their formal training. Many outside Stanford have commented admiringly on the huge advantage to the Stanford biochemists of Kornberg's foresight in establishing this collection.

more than simply trading bits of DNA between two physically distinct genomes. Genetic recombination is a fundamental biological process that faithfully restores the nucleotide sequence of recombined DNA molecules to their native state — with exquisite fidelity. This life-sustaining function was by no means novel to molecular biologists and geneticists. To be sure, the concept is as old as the field of genetics itself. No one had succeeded in mimicking accurate genetic recombination *in vitro*. But Berg wasn't bent on duplicating the exquisite reliability of genetic recombination in the test tube. He was contemplating the covalent *joining* of two genetically distinct DNA molecules on the test tube, regardless of their nucleotide sequence. Early adoption of the phrase *DNA joining technology* or *hybrid DNA technology* rather than *recombinant DNA technology* might have spared some ambiguity and misunderstanding in subsequent oral and written commentaries.

"We were aware of some important principles that already existed," Berg told an interviewer in early 1975. Notably, in the mid-1960s the celebrated Cold Spring Harbor Laboratory geneticist Alfred (Al) Hershey and his colleagues determined that the genome of bacteriophage λ is a linear double-stranded DNA molecule that terminates in overhanging single strands (so-called single-stranded tails) with complementary nucleotide sequence at each end of the molecule[1] (Fig. 2). As a consequence, the single-stranded ends can anneal and are accordingly often referred to as "cohesive" ends, or in more descriptive terms, as "sticky" or "velcro" ends. Consequently, cohesive ends can facilitate the circularization of one or more linear DNA molecules, or the formation of chains of two or more linear DNA molecules. To be sure, Dale Kaiser had earlier determined that the single-stranded cohesive ends of bacteriophage λ DNA are exactly 12 nucleotides in length.[2,3]

"Since the principle of joining DNA molecules through naturally-occurring complementary sequences was known, it became obvious that if we wanted to covalently join DNA molecules that are *not* naturally endowed with cohesive ends we would have to construct them biochemically," Berg related.[4] So he and his colleagues set about generating hybrid DNA molecules by deploying a series of sequential biochemical events now collectively referred to as *recombinant DNA technology*, a technology

Construction of Recombinant DNA

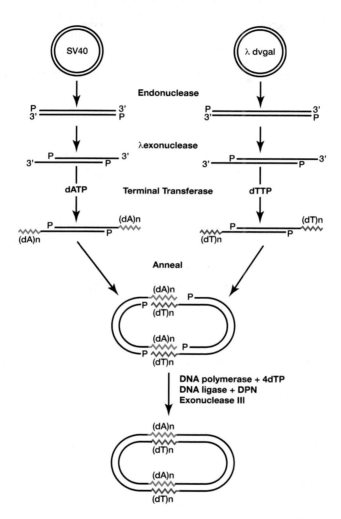

Fig. 2 Cartoon showing key steps in generating recombinants of SV40 and prokaryotic DNA molecules *in vitro*. (Courtesy of Paul Berg.)

which, as clarified in a later chapter, quickly matured to the capacity for cloning individual pieces of DNA of interest — even those containing just a single gene!

It dawned on the Stanford biochemist that a relatively straightforward way of generating DNA molecules with cohesive ends might rest in a well-known property of an enzyme called DNA nucleotidyltransferase, terminal transferase for short. Discovered by Fred Bollum and his collaborators in 1960, terminal transferase will catalyze the repetitive addition of any chosen nucleotide(s) to the termini of duplex DNA molecules, thereby lengthening single-stranded ends. Berg was familiar with terminal transferase. For one thing, as early as late 1967, he had followed studies by the German postdoctoral fellow Tom Jovin in Arthur Kornberg's laboratory, which had utilized the enzyme to make polynucleotides bound to cellulose. Additionally, Berg and his colleagues had previously used the enzyme to prepare polymers of known composition and sequence — so-called "block polymers."[5] Hence, it would be reasonable to conclude that he was certainly aware of the inviting properties of terminal transferase sometime in 1967, if not earlier.

Berg's ultimate plan was to exploit terminal transferase to attach a string of the nucleotides dTMP or dCMP to the single-stranded ends of linear duplex SV40 DNA, and a string of the complementary nucleotides dATP or dGTP to the "sticky" ends of linear phage λ DNA (Fig. 2). The expectation was that the artificially generated cohesive ends would spontaneously "velcro" to one another, thereby "recombining" two different DNAs. Such recombinant molecules would be designed to carry a specific gene(s) that would permit their identification and retrieval by appropriate genetic selection (Fig. 2).

The horde of stockpiled purified enzymes available in department freezers included a small amount of purified terminal transferase, a gift from F. N. Hayes at the Los Alamos National Laboratory (then called the Los Alamos Scientific Laboratory), who had previously donated the enzyme to Robert (Bob) Lehman, one of Berg's Stanford faculty colleagues. On January 16, 1970, Berg wrote to Hayes requesting more of the reagent.

> Sometime ago you were kind enough to send Dr. Lehman a preparation of your purified terminal transferase (or what you referred to as Addase). *Most of the enzyme* [suggesting that some of the enzyme was still available in the biochemistry department freezers] that you sent at that time has now been used up and I would like to prevail upon your generosity to see if we could beg another portion of the enzyme from you. In particular, we are setting up to do some experiments to try

to link two DNA segments together using the terminal transferase to generate complementary overlapping ends.[6]

Additionally, Arthur Kornberg had obtained a gift of the enzyme for Fred Bollum sometime in 1967, and conceivably some of that aliquot remained in the freezer. However, it is not certain whether Berg was aware of this acquisition.

A need for further innovation stemmed from the fact that the SV40 genome is naturally circular, i.e. there are no free ends (Fig. 2). Hence, Berg *et al.* had to devise a means of opening SV40 DNA circles (i.e. converting them to linear forms) in a manner that would yield unit length molecules bearing cohesive single-stranded ends of sufficient length to anneal two different DNA molecules. We'll return shortly to consider the crucial step of linearizing circular DNA molecules, which sets the stage for the series of biochemical events noted above.

The stability contributed by simply allowing two DNAs bearing single-stranded cohesive ends to anneal is a direct function of the length of the sticky ends. The longer the cohesive ends, the greater the stability conferred. This nuance was not appreciated in the Berg laboratory until Peter Lobban, a graduate student in Dale Kaiser's laboratory (see later), suggested that they increase the extent of the single-strandedness of one of the two termini generated by linearizing circular DNA molecules using another enzyme that attacks DNA called lambda exonuclease (Fig. 2). Only then could they deploy terminal transferase to render single-stranded ends cohesive. Ultimately, two different DNA molecules so generated would be coincubated in the hope of acquiring double-stranded conjoined (recombinant) DNA molecules possessing a certain degree of stability (Fig. 2).

The joining of two different DNA molecules as just described necessarily leaves the last nucleotide incorporated by terminal transferase sitting snuggly next to the following adjoining nucleotide, but not covalently joined to it. The significantly enhanced stability that derives by covalently joining these two nucleotides can be effected by an enzyme called DNA ligase. (The use of DNA polymerase together with an enzyme called exonuclease III shown in Fig. 2 are nuances that beg no explanation here to understand the big picture just painted. Lay readers might be further relieved to know that full comprehension of the series of biochemical reactions just described is by no means a prerequisite for comprehending the rest of the book.)

Believe it or not, this complex series of multiple successive enzymatic reactions actually worked! It is for that reason that the Stanford Department of Biochemistry went to the trouble of isolating and storing enzymes in anticipation of future experiments.

In a renewal proposal submitted in 1970 to extend his previously awarded research grant from the American Cancer Society, Berg wrote:

> [We] are exploring the feasibility of using SV40 DNA (or Py DNA) to transport non-viral genetic information into the cells they transform. It is our ultimate aim to attach a specific segment of DNA covalently by chemical or enzymatic means to the circular DNA molecule of either virus and then to determine (a) if cells can be transformed by such modified DNA's, (b) if the new genetic information is integrated with the viral DNA and (c) if they express the new information carried by the modified DNA.
>
> Our approach to synthesizing such potential transducing DNA molecules is as follows: (1) opening the viral rings and preparation of their ends for receiving the new segment; (2) isolation and preparation of the piece of DNA to be inserted; (3) joining of the two DNA's and ring closure to generate new, expanded circular DNA molecules.[7]

Berg set the experimental wheels in motion with two of his trainees. Postdoctoral fellow David Jackson was the first to initiate the construction of recombinant DNA molecules — in early 1969. When he entered Harvard College in 1960, Jackson was intent on majoring in English. But he was seduced by the biological sciences while attending a freshman course offered by the renowned educator and educational advocate (notably of women at Radcliffe), former Radcliffe president Mary Bunting-Smith.†

---

† Mary Ingraham Bunting (July 10, 1910–January 21, 1998) was an influential university president who was once profiled by *Time* magazine as its November 3, 1961 cover story. Bunting (or Bunting-Smith as she was sometimes referred to) became Radcliffe College's 5th president in 1960.

An accomplished microbial geneticist in her own right, Bunting-Smith's course was particularly popular because much of the content was provided by sophomores who had excelled as teachers in the course during their own freshman year — a privilege ultimately accorded to Jackson. "I was absolutely fascinated by the course and was motivated to work very hard during my freshman year," Jackson told an interviewer in 1978. "Bunting-Smith invited me to stay on and do summer research in her laboratory, and as a sophomore I joined the seminar as a speaker to the next group of freshmen."

David Jackson. (Courtesy of David Jackson.)

Jackson's decision to switch majors was conspicuously reinforced by exposure to another prominent Harvard figure in biology — James Watson, from whom he took a course in microbial genetics and virology as a junior. "Watson talked very, very softly," Jackson related. "So I used to come to class a good 10 minutes before the lecture to be sure of a front row seat. But what he said was just fantastic! He discussed scientific papers and thought about them in a way that I really hadn't been exposed to before."[8]

Watson was typically prompt for his 9:00 am lectures. But when on one October morning in 1963 he hadn't appeared by 9:30 am, someone in the class was dispatched to seek him out. "This classmate returned with a very odd look on his face," Jackson stated. "Without saying a word he walked up to the front blackboard and wrote: '*Dr. Watson has just won the Nobel prize!*' Watson rushed into class a few minutes later and spent the next hour or two telling us the story of what later turned out to be the book *The Double Helix!*"[9]

Jackson was secure in his conviction of becoming a molecular biologist well before he graduated from Harvard. He earned a graduate studentship in the department of biology at Stanford and identified Charles Yanofsky in the biology department as his PhD mentor. Berg, whom Jackson first encountered while taking his biochemistry course, also impressed him. "People working in the two labs were aware of Paul's and Charley's fervor about tennis and got to know each other quite well," Jackson commented. "So when I began thinking about a postdoctoral position I decided that the lab I most wanted to work in was Paul's."[10]

Jackson began his postdoctoral training in Berg's laboratory in March 1969. He spent part of his first year pursuing a project conceived while still in Yanofsky's laboratory. Meanwhile, Paul's trusty technician Marianne Dieckmann, together with postdoctoral fellow Francois Cuzin, (sometimes joined by Berg himself) initiated the experimental program soon after their return from the Salk Institute in the fall of 1968. Within a year the group was joined by an additional postdoctoral fellow, Donna Ardnt (soon to be Donna Arndt-Jovin) and subsequently by a new graduate student, John Morrow, who arrived in Berg's laboratory more or less contemporaneously with Jackson and by Janet Mertz, another MIT undergraduate.

Dieckmann, who worked with Berg for 43 years before she was struck down with terminal cancer, merits special mention. Born in Berlin, she was a young woman during the years of WWII. She was totally unsympathetic to the Nazi Party and refused to attend any of their propaganda camps. When the war ended, Berlin was devastated — and like many others, Marianne's primary concern was finding food to eat!

At war's end, she resided in East Berlin. Dieckmann yearned to attend medical school, but her request to leave East Berlin was flatly denied. Instead she became a qualified laboratory technician and left Germany for Boston, Massachusetts where she completed a medical technologist program at the Massachusetts General Hospital. Dieckmann was more than a mere technician who faithfully executed experiments designed by others. During this period she attended a lecture on DNA replication by

Berg in his laboratory with Marianne Dieckmann. (Courtesy of Paul Berg.)

Arthur Kornberg and was so taken with the experience that she promptly wrote to him asking whether she could join his laboratory as a technical assistant. "Arthur didn't need a technician, but he knew that I did," Berg stated.

"Marianne contacted me while I was still in St. Louis and I invited her to visit," Berg said. "When she arrived she telephoned and I arranged to pick her up from her hotel. She was astonished that I didn't resemble the stereotypic German 'Herr Professor'. Instead she concluded that I must be a mere post-doctoral fellow and expressed some petulance at being picked up by such an undistinguished individual! I hired her imme-diately — a smart decision, since she turned out to be nothing less than fantastic in the lab. She was amazingly quick at picking up anything new, especially techniques she had no previous experience with. After a while she even had her own research projects. When we moved to Stanford from St. Louis Marianne essentially managed all the required logistics, not only for my lab — but for the entire faculty group."[11]

Never married and with no children of her own, Dieckmann was mother to all, helping newly arrived students and postdoctoral fellows and their families to settle into life in Palo Alto and Stanford University. Forcefully but lovingly, she taught them the ethos and discipline required of a responsible and thoughtful scientist. She always knew how best to help students, reinforcing their confidence after mini-triumphs and providing solace in difficult times. Those far from home during the holiday season could always count on German-style Thanksgiving and Christmas dinners at Dieckmann's home.[12]

****

John Morrow was admitted to the graduate program in the Stanford Department of Biochemistry in the fall of 1969. Tall and lanky, Morrow was born and raised in Jackson, Tennessee. Some have commented that his demeanor is reminiscent of the stereotype of a Southern gentleman. After graduating from high school, Morrow was admitted to Yale University with a major in biophysics.[13] There, he became an ardent admirer of the late Frederic M. Richards, an accomplished biophysicist and crystallographer who was instrumental in the development of molecular biophysics and structural biology at Yale University — indeed nationwide.[14]

Morrow first encountered Berg when he visited Stanford to interview for admission to the biochemistry graduate program. "I was very interested in eukaryotic molecular biology, then an infant research field, and was drawn to the fact that Paul had just completed a sabbatical at the Salk Institute and was working with viruses," Morrow related. "I was convinced that the molecular biology of eukaryotic cells was going to be a hot topic of future research and I hoped to work with Paul even before I arrived at Stanford."[15] During a preliminary visit to Palo Alto, Morrow was hosted to dinner one evening by Peter Lobban, who as mentioned earlier, was a biochemistry graduate student in Dale Kaiser's laboratory. As related shortly, both Morrow and Lobban became central figures in the recombinant DNA story.

Though he and Berg had not yet defined a specific research project for a PhD thesis, Morrow began his bench work by familiarizing himself with growing polyoma (Py) virus in the laboratory. He found the switch to mammalian cells and viruses technically challenging. "Polyoma virus has a very extended life cycle under laboratory conditions," he stated. "It took literally days to grow and I spent many unproductive and frustrating days and nights in the lab, sometimes working alongside Donna and

Marianne Dieckmann." Jackson independently echoed these exasperated sentiments. "Working with this virus did not go well at first," he stated. "Trying to set up an animal cell culture operation in a biochemistry department, particularly where people were growing 100 liter vats of *B. subtilis* spores down the hall once or twice a week, was not easy. There were a lot of frustrations to contend with during the first year and a half or so after Paul got back from the Salk."[16]

While seeking respite from his technical frustrations with polyoma virus, Morrow learned that the closely related simian virus 40 (SV40) offered significant logistical advantages in the laboratory. "I told Paul that the polyoma stuff was not going well and that I knew where to procure SV40 virus and appropriate host tissue culture cells. With his acquiescence I introduced SV40 virus into the lab. This switch led to a significant gain in momentum and with the exception of Donna Arndt (who was firmly committed to an independent research product involving Py virus), everyone in the lab abandoned polyoma virus."[17]

Jackson's initial research project was also floundering, prompting Berg to begin contemplating a new direction for him. "One morning Paul and I got together and he suggested that we attempt to put new genes into SV40 DNA and use the recombinant molecules to introduce foreign DNA into animal cells," Jackson recalled. "I thought that carrying out detailed molecular biochemistry on how genes work in eukaryotes was hopeless at that time. But I also realized that one might possibly approach such a problem using a viral DNA, because it consisted of a block of genes that one could obtain in reasonable amounts and later use as a probe to measure transcription — and that sort of thing."[18]

Jackson enthusiastically launched the grand experiment. Sometime in 1970, he was joined at the bench by Robert (Bob) Symons, an experienced Australian biochemist on sabbatical leave from the University of Adelaide. After considerable struggle, Jackson and Symons ultimately achieved their experimental goals by deploying most (but significantly not all) the innovative biochemical maneuvers described earlier. As related more fully in a later chapter, in so doing the pair received helpful advice from Peter Lobban, who, in order to suit his own research purposes, had independently decided on an experimental strategy for generating recombinant DNA startlingly evocative to that devised by Berg. Also addressed more completely presently, Jackson and Symons were significantly assisted by a discovery by John Morrow in the course of his own research toward a PhD degree.

As mentioned above, the first task that faced Jackson and Symons was to find a means of converting circular SV40 DNA molecules to the linear state without chopping the DNA to countless fragments (Fig. 2). To achieve this elusive goal they employed a molecular scissor — an enzyme called an endonuclease that cuts duplex DNA. In living cells, DNA is a substrate for multiple such enzymes with different specificities, thereby subserving different biological functions. However, in the early 1970s, the only well-characterized endonuclease with which to conceivably achieve their objective was one called DNase I, an enzyme usually purified from extracts of bovine pancreas, but available commercially.

Previous studies by others designed to characterize purified DNAse I utilized Mg^{++} as the required divalent cation, under which conditions the enzyme catalyzed the generation of many single-strand breaks in duplex DNA — a property of no interest to Jackson and Symons, who sought DNA molecules containing a very limited number of *double strand breaks*, ideally just one.

Jackson resourcefully located a publication in the *Journal of Biological Chemistry* from David Goldthwait's laboratory (in none other than the Department of Biochemistry at Western Reserve University), which reported that in the presence of Mn^{++} (as opposed to Mg^{++}) the enzyme made *double-strand* breaks in duplex DNA.[19] But the daunting challenge remained of defining experimental conditions that on average yielded but a single double-strand cut could not be solved except by tediously examining all the products of incubation with DNAse I.

DNase I is not endowed with DNA sequence or base specificity; it cuts DNA at multiple random sites. The more enzyme added or the longer the incubation period the more cuts are generated. Indeed, incubation in the presence of DNase I can literally shred double-stranded DNA molecules to metaphorical ribbons! Establishing desired incubation conditions proved to be extremely time-consuming, inefficient — and disturbingly irreproducible.

Years later, Jackson reflected on this experience: "I spent most of the first year that I was working on the project learning how to open circular SV40 DNA with DNase I and obtaining a decent yield of full-length linear molecules. Furthermore, the biochemical procedures one had to go through before one could coax terminal transferase to generate single-stranded tails at an acceptable frequency was a whole lot more complicated than any of us realized."[20] Regardless, Berg remained convinced that the experimental strategy he and his colleagues were pursuing was technically feasible — and the grinding work continued.

Notes

1. Hershey AD, Burgi E. (1965) *PNAS* **53**: 325–328.
2. Strack H, Kaiser AD. (1965) *J Mol Biol* **12**: 36–49.
3. Wu R, Kaiser AD. (1967) *PNAS* **57**: 171–177.
4. PB interview by Rae Goodell, MIT Oral History Program, May (1975).
5. *Ibid.*
6. Letter to F. N. Hayes dated January 16, 1970. Reproduced with permission from the Department of Special Collections, Stanford University Library.
7. Research grant proposal to the American Cancer Society, dated October 14, 1970, p. 5, courtesy of Doogab Yi.
8. ECF phone interview with DJ, April 2012.
9. PJ interview by Lynnette Maloney, MIT Oral History Program, March 1978.
10. *Ibid.*
11. ECF interview with Paul Berg, January 2012.
12. *Ibid.*
13. ECF interview with John Morrow, September 2011.
14. http://www.csb.yale.edu/FMR/index.html
15. ECF interview with John Morrow, September 2011.
16. PJ interview by Lynnette Maloney, MIT Oral History Program, March 1978.
17. ECF interview with John Morrow, September 2011.
18. DJ interview by Lynnette Maloney, MIT Oral History Program, March 1978.
19. Melgar E, Goldthwait DA. (1968) *J Biol Chem* **243**: 4401–4408.
20. DJ interview by Lynnette Maloney, MIT Oral History Program, March 1978.

CHAPTER ELEVEN

Making Recombinant DNA; A Major Breakthrough

Berg brought his extensive expertise in nucleic acid biochemistry to bear

Having acquired stocks of SV40 virus and appropriate cells in which to propagate them, John Morrow turned his attention to identifying a research project that might yield a solid PhD thesis in a reasonable timeframe. "I was really struggling to get something going that would turn into a PhD thesis in a laboratory program that was at a very embryonic stage," he related.[1] Janet Mertz, who joined Berg's laboratory in the fall of 1970, about a year after Morrow, well remembers that "embryonic stage." "John had just brought SV40 into the lab, deciding that this virus might be better to work with than polyoma," Mertz related. "We were in the very early stages of getting things going; doing basic experiments such as working out a plaque assay to quantitate the yield of SV40 particles from infected cells — and things of that sort."[2]

An animated ("excitable" is an often used descriptor) and unabashedly vocal young woman whom Berg appropriately describes as "smart as hell," Mertz figures prominently in the recombinant DNA narrative, both with regard to her experimental contributions and in the origin of the controversy that soon centered around the new technology (see later chapters). Now a professor at the University of Wisconsin, Mertz, like her graduate student mentor, was raised in Brooklyn, New York. She attended the Bronx High School of Science, sacrificing many Saturdays to participate in the prestigious Columbia University Science Honors Program. As stated earlier, Mertz spent her undergraduate years at MIT, majoring in both biology and engineering. Significantly, she was witness to the radical socio-political gestalt prevalent in Cambridge, Mass. during the 1960s and early 1970s, a state of mind principally motivated by increasing public opposition to US military involvement in Southeast Asia. She was particularly sensitized to the spillover of these political sentiments to the emerging science of recombinant DNA technology, an effort viewed by some as yet another potential government-supported threat to the integrity and sanctity of biological species on the planet — including humankind.

A notable example of such political radicalization surfaced in late 1969, when Harvard University microbial geneticists James (Jim) Shapiro and Jonathan (Jon) Beckwith published a report in *Nature* announcing the isolation of a small fragment of bacterial DNA from a bacteriophage harbored in *E. coli* using genetic technology as opposed to biochemical approaches. Though clearly distinct from the potential that might materialize from recombinant DNA technology *in vitro*, the experiments reported by Shapiro and Beckwith indeed yielded fragments of double-stranded DNA enriched for the bacterial β-galactosidase (βgal) gene. These efforts offered the theoretical possibility of isolating a number of other bacterial genes in the same manner, though their potential was clearly confined to a mere handful that lent themselves to the experimental approach developed by the Harvard biologists.

Shapiro and Beckwith were vehement opponents of several aspects of government policy in general and the Vietnam conflict in particular, a mentality then possessed by many liberal academics. Very soon after their *Nature* report emerged, the pair announced a press conference, ostensibly to publically trumpet a method for isolating individual genes from bacteria by genetic maneuvers — though many scoffed at the newsworthiness of these experiments. In the course of the press conference the dreaded phrase "genetic engineering" crept into the conversation, raising the dramatic (to the press) specter of draconian consequences of future such experiments. As reported in a *Nature* editorial:

> A part of the trouble seems to have been a confrontation between the authors of the research and newspaper correspondents in Boston at the weekend; what seems to have caught the popular fancy is the awesome prospect of what might be done with genetic engineering if ever such a practice were possible.[3]

Regardless of the precise dialogue that transpired between Beckwith and Shapiro and a press contingent hungry for any hints of controversy, the two scientists blatantly underscored their own political sentiments in a letter to *Nature* published about a month later:

> In and of itself, our work is morally neutral — it can lead either to benefits or to dangers for mankind. But we are working in the United States in 1969 in a country that makes a prodigious use of science and technology to murder Vietnamese and poison the environment. The basic control over scientific work and its further development is in the hands of a few people at the head of large private institutions and at the top of government bureaucracies. These people have consistently exploited science for harmful purposes in order to increase their own power. The reality of the dangers we and others point out should not be minimized.[4]

This and other protests about the American presence in Southeast Asia were followed by a period of demonstration and intensive argument/counter-argument at MIT, Harvard and the city of Cambridge by protesters who identified with the rallying cry "science for the people." Such was the tenor of the socio-political environment to which the young and impressionable Janet Mertz was exposed. "So when I went out to Stanford I had in the back of my mind this recent thing concerning the Shapiro–Beckwith experiment and the controversy that resulted and so forth, and I pretty much held the radical point of view that we scientists shouldn't go around developing genetic engineering," she stated. Though never a political activist in her own right, upon arriving on the Stanford campus

Janet Mertz. (www.mcardle.wisc.edu/faculty/bio/mertz.j)

in the fall of 1970, Mertz was surprised by the generally indifferent views expressed by the academic community there, especially students. "During 1969 up to the summer of 1970 there were riots in the Boston area on an almost weekly basis," she commented. "Things didn't start getting rough at Stanford until later."[5]

In addition to her political "radicalization" at MIT, Mertz arrived at Stanford at a time when American women were not actively encouraged to pursue graduate study. Indeed, she was but the second female graduate student in the Stanford biochemistry department in the previous eight years. One prominent academic institution at which Mertz interviewed unabashedly expressed anti-feminist sentiments. Then too, she was not yet rid of her apprehension about the ethical and moral implications of genetic engineering in general and was somewhat concerned to discover that Berg and others in his laboratory were actively gearing up to launch recombinant DNA technology.[6]

****

Anti-war activism on the Stanford campus from the early days of the Vietnam conflict was indeed late in coming. But protests increased in frequency and intensity as the war entered the decade of the 1970s. Indeed, the Vietnam era showcased a record high of Stanford student activism, from protests to sit-ins to loud and emotional anti-war rallies. The 1970s also proved to be a crucial time of change for the university itself, with numerous policy changes, including the replacement of university president Ken Pitzer by provost Richard Lyman, and as mentioned previously, removal of the "politically incorrect" Indian mascot of the Stanford athletic program and even changing the athletic program moniker — "the Stanford Indians" to the bland "Stanford Cardinal."

Berg recalls a grim atmosphere at Stanford during the Vietnam era. "For people of my generation — born long before 1971 — it was difficult to imagine just how different the mood was at American universities. Students staged marches and sit-ins constantly, often with faculty in the lead — and the threat of violence hung constantly in the air. The previous year four students had been shot dead by National Guardsmen on the campus of Kent State University in Ohio while protesting the US bombing of Cambodia."[7] In the spring of 1971, the spark at Stanford that heralded imminent trouble was civil rights. A black custodian at the medical school had been fired after raising a stink about racial disparities in salaries. The medical school had also denied tenure to a Latino faculty member."[8] This and subsequent events were graphically described in an article published electronically under the title "The Stanford Medical Center Protest: Looking for Trouble":

> By the early 1970s, students and leftists were looking for new causes to pursue. This was the scene as students and leftist activists focused on a small but real inequity at Stanford University Medical Center — the salary of the custodians. The condition of the working poor was the sort of issue that galvanized students and baffled the establishment. While students saw it as a good place to think globally and act locally, hospital administrators were caught off guard by the sudden and unexpected outrage.
>
> The controversy began when custodian Sam Bridges was fired in late March of 1971. Bridges had been trying to organize custodians at the hospital and discussing the merits of Black Power. His dismissal led to condemnation among civil rights and student groups on campus. On April 8th, 1971, 250 protesters marched to the office of hospital

director Dr. Thomas Gonda to further press demands. Finding him out to lunch, 60 of the protesters decided to stage an overnight sit-in, occupying administration offices between the hospital's surgical clinic and blood bank. The activists included members of campus groups and outsiders, including the Black United Front, the Black Liberation Front, Venceremos and the Latin Alliance. Demands were listed that encompassed both the Bridges firing as well as another perceived injustice at the hospital — the refusal to grant tenure to Latino doctor Jose Aguilar.

The next day as negotiations stalled at the sit-in's 30 hour mark, Stanford officials called the Palo Alto Police Department. Arriving in riot gear, some 70 Palo Alto cops and nearly 100 sheriff deputies congregated outside the hospital offices along with impartial observers and local press. At 5:55 PM, Assistant Police Chief Clarence Anderson declared the sit-in to be an "unlawful assembly" and gave the protesters five minutes to disperse. When they did not, the police prepared to forcefully clear the area. What followed was a violent and chaotic melee.

In response to these perceived injustices, students and outside activists occupied the Stanford Medical Center. Some 36 hours later, after repeated attempts at negotiating the occupants' departure, police stormed the building. According to a report in the *Stanford Daily* (the student daily newspaper), "protestors had smashed all the furniture and used chair legs as clubs. Riot police had to resort to mace; twenty-three demonstrators were arrested; twenty people suffered injuries. The hospital suffered $100,000 in damages. 'It looked like a war zone,' the dean of the medical school stated. 'Every window that could be reached by a stone was broken.' "[9]

In a separate incident, the Black Panthers (members of an African-American revolutionary leftist organization from the mid-1960s to the early 1980s) stormed into a meeting of the Medical School Executive Committee, a committee of department chairs led by the dean of the medical school that met on a regular basis. The intruders "ringed the room, arms folded, waiting to hear the response to their demands regarding workers' rights and affirmative action. The assembled group was unambiguously intimidated. 'I wondered if I would make it home that night,' one committee member reputedly stated. However, the late Henry Kaplan, then head of the Stanford Department of Radiology and an internationally recognized innovator of radiation therapy, agreed to represent the committee in dialogues with the medical school Black Advisory Committee and with a Latino group called Alianza Latino."[10]

In keeping with the general mood of the time, Berg acceded to an unusual and ultimately historic request from a group of enterprising graduate students. "In the summer of 1971 three graduate students knocked on my door," he related. "They informed me that they were intrigued by my lectures on protein synthesis and that they and their ballet dancer girl friends wished to capture the essence of that process in dance. At the time I thought they were nuts — or simply kidding me! But they didn't need much money so I agreed to provide funding for the filmmaking. It seemed to me too that the musical event they had in mind might soften the generally disquieting mood then prevalent on the campus."[11]

The November 5, 2010 issue of *SCIENCE* dug back in history and reported this groundbreaking event at Stanford orchestrated by Berg and his students:

> The result, shot in a single day with little rehearsal, is deeply trippy. It begins with a 3-minute introduction by Berg in front of a chalkboard. "Only rarely is there an opportunity to participate in a molecular happening," intones Berg without emotion. He runs through a play-by-play of protein synthesis. Then the dance begins. It involved so many dancers that the film crew had to put the camera on the roof of one of the tallest buildings on campus and shoot from above.
>
> The dance begins with a pile of 20 students in blue hospital gowns — representing the 30S subunit of the ribosome — oozing across the grass. Then a hand-linked line of about 50 people with colored balloons attached to their heads — representing the strand of mRNA — snakes into the action. An even larger crowd in blue — the 50S subunit — then somersaults into place. Colorfully dressed women (the amino acids) kidnapped by merry gangs (the tRNAs) come bounding in. The two crowds of subunits writhe and bow in unison as the mRNA crowd slides between them.

Adding to the surrealism, there's a live band jamming, with a singer belting out lyrics like "Ohhhh hydrogen bond! Woo hoo!" Meanwhile, a woman softly recites a modified version of Lewis Carroll's *Jabberwocky* that explains the molecular mechanisms being danced.

Besides giving the intro lecture, Berg's only contribution to the choreography was the concept of energy. "Every peptide bond requires consuming energy and the movement of the mRNA relative to the ribosome requires energy," Berg related. "I came up with the idea of using a fire extinguisher. A sinister-looking young man in a red cloak and goggles represents the energy-providing molecule GTP. You'll see that the guy goes into the middle of the ribosome, and suddenly up comes a puff of smoke," says Berg. A high-definition version of this video can be viewed online.

SCIENCE interviewed Paul Berg, as well as David Thomas, one of the Ph.D. students who took part. "Nobody who participated in that dance remembers anything about it, due to pharmacological events out of our control," said Thomas, now a biophysics professor at the University of Minnesota. "It's a good thing they recorded it."[12]

In 2008, *SCIENCE* initiated a revival of the precedent set at Stanford by sponsoring an annual event around the country called "Dance Your Ph.D." The rules were simple, essentially: "Using no words or images, interpret your Ph.D. thesis in dance form." Entrants were divided into three categories — graduate student, postdoc, and professor — and the prize for each was a year's subscription to *SCIENCE*.[13] By 2010, additional sponsorship was offered by an entity called TedXBrussels launched by Steve Wozniak (who together with Steve Jobs and Ronald Wayne co-founded Apple Computer). TedXBrussels offered a free trip to Brussels for the winning dance!

The Stanford convulsions were by no means confined to students. In May 1969, the *New York Times* broke the news about the secret bombing in Cambodia authorized by President Richard Nixon. Stanford, along with a number of other universities around the country dispatched faculty delegates to Washington to protest this action. Berg was invited to join the Stanford delegation. "I don't recall how or why I would have been identified as a spokesperson for the university," he stated, "except for the fact that I had a pretty good rapport with the medical students at the time, possibly because I had been involved with some of the so-called teach-ins intended to educate people to the horrendous events in South-East Asia."[14]

Berg has no recollection of a protest trip to Washington, but he accepted the reality of this occasion when confronted with a letter dated August 11, 1976 from an individual stating: "I was at Stanford from 1966 to 1970. In the spring of 1970 you and I shared a room in Washington, D.C. as part of the Stanford delegation to protest the invasion of Cambodia."[15]

His activist sentiments fully aroused, a week later Berg wrote an impassioned letter to Charles H. Percy, a Republican US Senator from Illinois, who remained in office for nearly 20 years and who devoted much of energy to business issues and foreign relations:

> I wanted to thank you deeply for the time you spent talking with us. Although I am not of your voting constituency, and not even of your political party, I believe that you share our deep concerns for what is happening in our country.
>
> In my view, there can be no, I repeat no, justification for perpetuating an action in the name of our country when that action is ripping us apart. The cost of human lives (and the Third World Group tried to tell you that the loss of Asian lives is just as grievous as the loss of American lives) is tragic enough. The crippling effects on our economy (why is it not seen more clearly that the inflation that mocks our progress is due primarily to the outrageous costs of maintaining the war?) and the inability to address [us] constructively to the Vietnam adventure have become serious issues.
>
> ··· The slogan "Power to the People" may have ominous meanings to many, but to me it means power to the people's representatives, the Congressman and Senators. If they fail to exercise their constitutional responsibility and turn their backs on their legally based prerogatives, we shall see that the term "law and order" has little meaning except as a lever for repressing political dissent or protecting the status quo.[16]

Berg sent a modified copy of this letter to another Republican Congressman, Paul N. McCloskey. And while in this indignant state of mind he composed yet another epistle, this to the popular TV talk host Dick Cavett, then with ABC TV:

> When watching the broadcast in which you interviewed President Nixon's Director of Communications, Herbert Klein, I thought your questions were penetrating, provocative and spoke to the issue; they served to expose the bankruptcy of the Administration's policy and actions in Southeast Asia in general and in the Cambodian venture in particular.[17]

The remainder of this letter vehemently challenged the accuracy of comments attributed to Nixon about the student "bums" who allegedly set fire to the Center for Advanced Study of the Behavioral Sciences at Stanford University. Years later, Berg took on the *National Review*, the respected monthly periodical edited by the conservative intellectual William F. Buckley. Berg had been asked by the magazine to write a commentary to accompany an article on the recombinant DNA controversy by a scientist who shall remain nameless, but whose opinions Berg didn't much care for, prompting a disparaging letter to the *National Review*.

> Although I am not a frequent reader of the *National Review* I have always viewed it as a magazine committed to responsible journalism. The *National Review* exudes probity, integrity and responsibility and pandering to sensationalism has been left to others. But your recent letter telling me of your intention to publish the "piece" by ·············· entitled *The Recombinant DNA Controversy* astonished me even more than the outrageous distortions the article contains. Having heard William Buckley demolish flimsy, stupid and self-serving arguments in the past, I'm at a loss to understand how he, as Editor-in-Chief, could accept these feckless arguments in his own magazine. ··········[18]

Though not obviously "red hot" politically, as just noted Berg possesses definite political views and may well have been identified as an "ultraliberal" at Stanford during that period. Aside from his activism over the conflict in Southeast Asia, at a time when protest about racial discrimination was rampant in the United States, Berg and a group of faculty colleagues enterprisingly organized a summer course for black students from one of the local high schools. "Essentially no students at that school went to college." Berg related. "We applied for and received a National Science Foundation (NSF) grant to launch a summer program for a select group of these students. We had about 25–30 of them, all identified by teachers at the school as underperformers. We arranged for the students to be bussed to the medical school and we provided lunch. The kids worked in a lab doing simple exercises. For example, I would crumble a yeast cake, add a drop of water, put some on a slide and ask them to describe what they saw under the microscope. We also armed the students with Petri plates and swabs and instructed them to take swabs from anywhere in the medical school and see what grew from them. The class convened every week day for about eight weeks of the summer."[19]

Berg (front row, second from *left*) with his class of ethnic minority high school students. (Courtesy of Paul Berg.)

"For several years the course enjoyed enormous enthusiasm and support from all quarters," Berg continued. "But eventually some of the more militant parents, mainly mothers, developed a (totally false) perception that money to support the program was enriching the coffers of Stanford University rather than the program itself. This belligerent group of mothers began picketing in the courtyard of the medical school and kids who really wanted to attend classes had to cross a picket line. Eventually the program was dropped."[20]

****

In the run up to the time that Janet Mertz chose to work in Berg's laboratory, the pair had enjoyed several extended conversations about ongoing research in his laboratory, always with a view to identifying elements of the program that might be both attractive to her as PhD thesis projects — and mollify her social conscience. In the end Mertz rationalized that at least for the time being she would "go ahead and work on the research problems discussed, on the grounds that people would probably reason that being able to put some DNA into [a simple bacterium such as *E. coli*] is a long way from being able to change people's genes."[21]

"One of the projects that Berg suggested to me was to achieve the converse of what Dave Jackson was attempting," Mertz elaborated. "At the time that Dave was working on the development of recombinant DNA techniques the idea was to take a piece of bacterial DNA, such as the entire galactose operon of *E. coli*, or another DNA that we had some understanding about, hook it up to SV40 DNA and transform animal cells with the recombinant molecules. One could then try to examine expression of DNA fragments containing defined bacterial genes in animal cells. Berg suggested a complementary experiment to me, namely putting SV40 DNA into *E. coli* so that one could also study expression of eukaryotic DNA in bacteria."[22]

Mertz thought long and hard about her discussions with Berg. She was unequivocally delighted to be exposed to the academic milieu of Stanford University (not to mention the year round mild Northern California weather) and was keen on pursuing research under Berg's mentorship. Besides, trying to express recombinant DNA was indeed a long way from introducing foreign genes into human beings, and Berg and others in the lab kept reassuring her that her fears were groundless. "Paul had just written an American Cancer Society grant proposal in which he was proposing to make recombinant DNAs that could be studied in both mammalian cells and bacteria. I thought about the project over Thanksgiving (in 1970) while visiting relatives in Los Angeles and decided to join Paul's group shortly thereafter." In January 1971, Mertz donned a laboratory coat and got to work!

Janet Mertz was by no means a novice in a genetics/molecular biology laboratory. She had been exposed to a hefty dose of bacteriophage genetics under Ethan Signer's tutelage at MIT and felt perfectly capable of handling the project Berg had in mind. But she also keenly appreciated that before any of these potentially groundbreaking experiments could be attempted a number of preliminary logistical issues required solution.

"To get started Paul suggested that I become familiar with how to infect mammalian host cells with SV40 and to quantify viral titers using the standard plaque assay," Mertz related. "So one of the several projects that I initiated was to optimize conditions for infecting cells with SV40 virus. He also suggested that I compare the infectivity of circular and linear SV40 DNA."[23] An important outcome of those experiments is related later.

At about the same time Mertz addressed the issue of which bacterial DNA might be "recombined" with SV40 DNA. The availability of a strain of λ bacteriophage harboring the entire *E. coli* galactose operon seemed like a reasonable source of bacterial genes, and would also provide a convenient genetic marker for identifying and isolating recombinant DNA molecules.

Sometime in the spring of 1971, Mertz teamed up with David Jackson and Douglas Berg (a postdoctoral fellow in Dale Kaiser's laboratory — no relation to Paul), who were then following a procedure developed some years earlier by Kenichi Matsubara in Dale Kaiser's laboratory. Matsubara produced a deletion mutant of phage λ that replicated stably in *E. coli* as a small circular plasmid. He referred to the plasmid as λ*dv*. In 1971, Doug Berg created a similar plasmid called λ*gal*. A derivative plasmid was designated λ*dvgal* because it contained both the deletions in λ*dv* and the complete set of genes that *E. coli* uses to metabolize the sugar galactose. Jackson and Mertz established that λ*dvgal* is a small circular double strand DNA with the capacity to replicate in *E. coli* as an autonomous "mini-chromosome."

Mertz was quick to recognize that λ*dvgal* might conceivably serve as an appropriate vector for carrying SV40 genes into *E. coli*. She also appreciated the potential utility of the *gal* gene for selecting rare bacterial cells that had acquired SV40 DNA. On the other end of the Berg laboratory, the availability of λ*dvgal* afforded Jackson a defined segment of the *E. coli* genome that he might possibly link to SV40 DNA and introduce into animal cells in order to examine expression of bacterial genes in such cells. Construction of the desired phage vector and preliminary transfection experiments progressed favorably, though Doug Berg, Jackson and Mertz did not get around to publishing the final construction and testing of λ*dvgal* until late in 1974.[24]

Lacking experience with growing and maintaining mammalian cells in culture, Mertz was excited to learn about a three-week course to be offered in the early summer of 1972 at the Cold Spring Harbor Laboratory on Long Island, New York — a course designed to train novices in the safe handling of mammalian cells and some of the viruses that infect them. The attendant opportunity to visit her parents in New York provided an additional incentive to return to the East Coast.

The course was not designed for technically naive graduate students like Mertz. It was primarily intended to meet the needs of established investigators keen on launching research programs involving the use of eukaryotes. Never shy in the pursuit of scientific goals, Mertz lobbied Berg for permission and the required financial support to attend the course. We'll consider the far-reaching political implications of Mertz's participation in the Cold Spring Harbor course in a later chapter (see Chapter 17).

****

Given the difficulties associated with using DNaseI to generate unit length linear SV40 DNA, the Berg group was constantly on the lookout for a more amenable endonuclease. Since John Morrow was still trying to identify a project suitable for his PhD thesis Berg suggested that he set about establishing a physical map of the SV40 genome as his PhD thesis research project, an objective that had been in Berg's mind well before Morrow joined the laboratory. "Before I left for my sabbatical Dale Kaiser and Dave Hogness were trying to map phage λ genes," Berg related. "They had developed a DNA transfection assay which enabled them to introduce segments of λ DNA generated by high-speed shearing into *E. coli*, and determine which phage genes they contained and where in the phage genome they came from. In their case the natural ends of the λ DNA molecules served as the obligatory fixed point of reference essential to their mapping studies."[26]

Generating a physical map of the SV40 genome was obviously confounded by the fact that the DNA was a covalently closed circular structure — without a fixed reference point. Berg was aware of reports by Daniel (Dan) Nathans and Hamilton (Ham) Smith at Johns Hopkins University who had used a restriction enzyme (called *Hin*dII/*Hin*dIII isolated from the bacterium *Hemophilus influenza*) to cut the SV40 genome into a number of discrete fragments that could be resolved by gel electrophoresis. Coincidentally, by hybridizing individual DNA fragments to mRNA from SV40-infected cells, they demonstrated that some of the fragments contained genes that were expressed early after infection, while others carried so-called late expressed genes. Control of the regulatory switch from early to late expression of SV40 genes subsequently became a focus of considerable interest in the phage λ community in general.

But Berg was in search of a restriction endonuclease* that would generate a single (or a very limited number of breaks) at a unique site in SV40 circular DNA, thereby establishing a fixed point to serve as a physical reference for genome mapping. Soon after returning from his sabbatical in La Jolla, he had acquired several restriction endonucleases produced from *E. coli* as a gift from Matthew (Matt) Meselson at Harvard to hopefully assist the visiting Francois Cuzin in early experimental endeavors. But to everyone's disappointment, the enzymes obtained from Meselson cut SV40 DNA at multiple sites. After Cuzin left Berg's lab to return to France, the mapping studies were pursued by John Morrow. He scoured the world for any and all new restriction endonucleases he could lay his

---

* As the name implies, restriction endonucleases, widely distributed in nature, evolved to "restrict" the propagation of unwanted foreign DNA molecules that might enter bacterial cells.

John Morrow. (Courtesy of John Morrow.)

hands on. But like Cuzin before him, Morrow was frustrated to discover that none of them yielded a unique single cut in SV40 DNA.

Notes

1. ECF interview with John Morrow, September 2011.
2. ECF interview with Janet Mertz, February 2012.
3. *Nature* **224**: 834–835, 1969.
4. *Nature* **224**: 1337, 1969.
5. JM interview by Rae Goodell, MIT Oral History Program, March 1977.
6. ECF interview with Janet Mertz, February 2012.
7. ECF interview with Paul Berg, June 2012.
8. *Ibid.*
9. http://www.paloaltohistory.com/the-stanford-medical-center-protest.php
10. Jacobs CD. (2010) *Henry Kaplan and the Story of Hodgkins Disease.* Stanford University Press, p. 268.
11. ECF interview with Paul Berg. June, 2012.
12. Bohannan J. (2010) Why Do Scientists Dance? *SCIENCE* **330**: 752.
13. http://www.sciencemag.org/content/319/5865/905.2.full
14. ECF interview with Paul Berg, February 2012.
15. Letter from Patrick Shea, August 11, 1976. Reproduced with permission from the Special Collections, Stanford University Libraries.
16. Letter to Charles H. Percy, May 11, 1970. Reproduced with permission from the Department of Special Collections, Stanford University Libraries.

17. Letter to Dick Cavett, May 11, 1970. Reproduced with permission from the Department of Special Collections, Stanford University Libraries.
18. Letter to Kevin Lynch, *National Review*, June 27, 1978. Reproduced with permission from the Department of Special Collections, Stanford University Libraries.
19. ECF interview with Paul Berg, February 2012.
20. *Ibid.*
21. JM interview by Rae Goodell, MIT Oral History Program, March 1977.
22. *Ibid.*
23. ECF interview with Janet Mertz, February 2012.
24. Berg P, Jackson DA, Mertz JE. (1974) *Virology* **14**: 1063–1069.
25. JM interview by Rae Goodell, MIT Oral History Program, March 1977.
26. Interview with Paul Berg, June 2012.

CHAPTER TWELVE

*Eco*RI Restriction Endonuclease — A Major Breakthrough

Therefore, any two DNA molecules with RI sites can be recombined at their restriction sites

In the midst of his frustration about generating a physical map of the SV40 genome, John Morrow was informed by Berg about a new restriction enzyme from *E. coli* recently discovered in Herbert (Herb) Boyer's laboratory at the nearby University of California in San Francisco (UCSF).

Boyer had become interested in restriction endonucleases in the mid-1960s when he learned that DNA restriction and modification enzymes interacted with the same sites in DNA and reasoned that "this would be a great way to study site-specific interactions between proteins and DNA."[1] After plowing through a number of available enzymes, he disconsolately concluded that this goal was easier said than done with existing technology. But while reading the literature, he learned that scientists in Japan had discovered that bacteria carrying antibiotic resistance factors often possessed enzymes for the restriction and modification of DNA from other species. So he dropped thoughts of site-specific interactions between proteins and DNA and elected to investigate these newly identified enzymes and their possible relationship to antibiotic resistance.

"I had a graduate student with a degree in clinical microbiology and had experience working in a hospital medical microbiology lab," Boyer stated, "a wonderful guy named Bob Yoshimori."[2] Boyer had dispatched

145

Yoshimori to the clinical laboratory at the UCSF hospital with instructions to obtain as many cultures of drug-resistant *E. coli* as possible. Yoshimori returned with an impressive haul of isolates, a number of which possessed restriction and modification activity. Among them was a restriction enzyme that was dubbed *Eco*R1, isolated from a woman who had been admitted to hospital with a urinary tract infection resistant to multiple antibiotics. Remarkably, *Eco*R1 enzyme was not found anywhere else for another 15 years — when it was located in a freshwater microorganism.[3]

Berg contacted Boyer on Morrow's behalf late in the summer of 1971. The UCSF investigator generously agreed to provide some *Eco*RI enzyme, prompting Morrow to jump into his car and dash up to San Francisco. Upon introducing himself and acquiring some enzyme (enough to last a lifetime, according to Boyer) Morrow was thrilled to be invited to lunch with Boyer and his colleague Harvey Eisen. As noted later, Morrow's acquaintance with Boyer was to have significant consequences.

On returning to Stanford with his precious acquisition, Morrow latched on to a seminal observation. *Eco*RI endonuclease cut SV40 DNA at a single unique site! At long last he had the much wanted reference point in the SV40 genome for mapping studies, work that would be embellished with numerous other restriction enzymes and would ultimately lead to his PhD thesis. Aside from this auspicious breakthrough, Morrow's observation coincidentally provided Jackson and Symons with a more facile way of opening circular SV40 DNA for their recombinant DNA experiments — and they immediately abandoned using DNase I to generate unit length linear DNA. Indeed, when Jackson, Symons and Berg published their decisive paper on the generation of recombinant DNA in the October 1972 issue of the *PNAS*, they didn't even bother to mention the use of DNase I. Such is the weird and wonderful role of serendipity in science.

This key paper, the first documenting the formation of making recombinant molecules in the laboratory carrying both SV40 and bacterial DNA, bore the extended title: "Biochemical Method for Inserting New Genetic Information into DNA of Simian Virus 40: Circular SV40 DNA Molecules Containing Lambda Phage Genes and the Galactose

Operon of *Escherichia coli*." The manuscript was submitted in late July 1972, and as just mentioned, was published in October of that year.[4] On August 16, 1972, a mere two weeks after submitting this manuscript to the journal, Berg, acting again in his capacity as a member of the US National Academy of Sciences, dispatched a second manuscript to the journal entitled "Cleavage of Simian Virus 40 DNA at a Unique Site by a Bacterial Restriction Enzyme," in which he and Morrow reported Morrow's new and crucial observation.[5]

Toward the end of the summer of 1971, Janet Mertz returned to Stanford from the Cold Spring Harbor course and a brief stay with her parents in New York. Still concerned about the perceived moral and ethical implications of genetic engineering in general, she was alternately relieved and disappointed to discover that facing the same sort of dilemma, Berg had unilaterally decided to impose a moratorium on any further recombinant DNA experiments in his laboratory that involved the introduction of SV40 DNA into *E. coli*. We'll return to a consideration of the events that led to this drastic decision presently.

"Paul was professor and chair of the biochemistry department," Mertz proclaimed, "Whereas I was merely his lowly second-year graduate student. It really didn't matter whether or not I thought the moratorium was appropriate or silly. Nobody asked my opinion! So I had essentially no choice but to go along with it, which in fact only forbade one series of experiments — introducing SV40 joined to λ*dvgal* into *E. coli*, which, based on my earlier discussions with Paul, I had been planning to do for my PhD thesis. So after the Cold Spring Harbor course in 1971 I dutifully discontinued recombinant DNA work."[6]

Mertz's laboratory notebooks document that during 1971–72, she was more or less simultaneously examining the infectivity of different topological forms of SV40 DNA, characterizing naturally arising defective SV40 genomes, and developing methods for isolating and propagating individual clones of SV40 mutants, expertise sorely needed in the Berg laboratory.

Following Berg's suggestion that she examine the infectivity of different forms of SV40 DNA, Mertz began testing the viability of double-stranded linear SV40 DNA generated with *EcoRI* endonuclease. Upon her return to Stanford, she continued these studies and consistently observed residual viable viruses — at a frequency of about 5–10% that of uncut DNA. Since linearized SV40 DNA was known not to be infective, this puzzling result yielded no obvious explanation.

Berg's initial reaction to Janet's results was to conclude that she had not purified the linear DNA sufficiently to be rid of all circular DNA, hence the observation that 5–10% of the DNA yielded viable phage. At his suggestion she repeatedly purified *Eco*RI linear DNA by ultracentrifugation — but continued to observe residual infectivity. At this point, Berg recommended that Mertz consult with Ron Davis, an accomplished and experienced biochemist and DNA electron microscopist. Electron microscopy confirmed that Mertz's DNA preparations did not contain residual circular DNA in amounts that could account for the infectivity.

Mertz and Davis reasonably dismissed the unlikely possibility that some fraction of the SV40 DNA was resistant to *Eco*RI endonuclease. Instead they considered that the enzyme might have generated DNA with short cohesive ends that could spontaneously reanneal at a low frequency to reconstitute a fraction of circular molecules that yielded plaques on infection. Further head scratching yielded the more cogent hypothesis that *Eco*RI enzyme was indeed cutting circular SV40 DNA efficiently. But if for some reason(s) the DNA ends were not covalently joined to extant DNA, only a small percentage of recircularized DNA molecules would be observed at any given time. "I had no intention of turning this observation into a full-scale research project," Mertz related. "But I decided to ask, perhaps as a last-ditch experiment, whether I could increase the frequency of circularized molecules by simple DNA ligation, a result anticipated if incubation of DNA linearized with *Eco*RI generated cohesive ends in molecules that didn't circularize because they were not stably ligated."[7]

Mertz obtained purified DNA ligase from Paul Modrich, a graduate student colleague in the biochemistry department in Bob Lehman's laboratory. As she and Davis considered the details of the new planned experiments, they realized that they had no idea how long (or short) putative cohesive ends in SV40 DNA generated by *Eco*RI enzyme might be.

If Mertz incubated cut SV40 DNA in the presence of DNA ligase in the wrong temperature range, cohesive ends might indeed not be stably ligatable. Ultimately, she established incubations at 15°C before examining the DNA under the electron microscope. "Lo and behold, I had circles," she exclaimed. "I was ecstatic and loudly yelled, 'Yahoo!!' We had optimal conditions the first time we tried. Ninety-eight percent of the DNA had reformed into circles!!"[8]

Davis took the experiments a step forward. Since electron microscopy had unequivocally demonstrated that Mertz had generated essentially pure linear DNA, he suggested mounting the DNA on the electron microscope grid in the cold room. Under those conditions, the majority of the molecules were circular. But if he raised the temperature at which the DNA was mounted, the pair observed increasing numbers of linear molecules. Davis was able to estimate the melting temperature (T_M) of the transition at about 3–4°C, an observation indicating that the cohesive ends were short. Additionally, he predicted that these short sticky ends were likely exclusively composed of A paired with T (less stable than G:C base pairs), a prediction that turned out to be correct once the cohesive ends were sequenced by Herb Boyer and his colleagues. Mertz's concern about the temperature at which to ligate ends was clearly warranted!

Mertz and Davis extended these experiments to show that all the SV40 cohesive ends were the same length, later revealed by DNA sequencing to be just four nucleotides. And "as a final experiment, to show that it was a general phenomenon that worked with any DNA, we mixed SV40 and λ*dvgal* and showed that we could join linear bacterial and eukaryotic DNA," Mertz explained.[9] In short, while Mertz and Davis had not set out with the specific intention of demonstrating that *EcoRI* enzyme and DNA ligase could catalyze both the linearization of circular DNA and the covalent joining of two different linearized DNA molecules to yield recombinant DNA — that is precisely what they had achieved! "These chimeric molecules would have been able to replicate in *E. coli*," Mertz and Berg wrote.[10]

That supposition was not tested experimentally consequent to Berg's self-imposed moratorium on generating *E. coli* cells that might contain

Ron Davis. (med.standford.edu/profiles/Ronald_Davis/)

SV40 oncogenes. As related more fully later, during a formal discussion at the Cold Spring Harbor course, Mertz innocently revealed the essential nature of the studies in progress at Stanford. It was then common knowledge that SV40 virus carries genes that are oncogenic in mice. Hence, the experiments communicated by Mertz sparked a wave of anxiety in some of the course instructors that if humans became infected with *E. coli* carrying SV40 virus, an epidemic of human cancer might conceivably transpire! So began the so-called *recombinant DNA controversy*, a historic saga fully recounted in Chapters 16–23.

Mertz's and Davis' studies transpired at roughly the same time that Berg was writing the manuscript that — as just mentioned — he, Jackson and Symons submitted to the *PNAS* in the summer of 1972. He was then also in the midst of composing a second manuscript reporting Morrow's startling observation that *Eco*R1 endonuclease cuts SV40 DNA at a single unique site. Approximately three weeks after contributing this manuscript for publication, Berg communicated a third manuscript to the journal announcing Mertz's and Davis' observations that incision of DNA with *Eco*RI yields cohesive ends. At that time (and possibly to this day), the journal did not permit a single individual to author more than two papers in any one issue of the journal. Hence, Mertz and Davis could not identify Berg as a co-author on their paper. In fact, this important publication is not even listed in Berg's formal bibliography!

Notably, with regard to later events concerning priority for this observation, the abstract of the published paper prophetically stated: "*Therefore, any two DNA molecules with RI sites can be recombined at their restriction sites by the sequential action of RI endonuclease and DNA ligase to generate hybrid DNA molecules* [author's italics]." The discussion section went even further. "·········· *quite possibly with appropriately chosen concentrations and molecular species of DNA one may, in this simple way, be able to generate specifically oriented recombinant DNA molecules in vitro* [author's italics]." While the phrases "gene (or DNA) cloning" were not explicitly offered, such predictions are clearly implicit in the statements just quoted. Bluntly stated, though the Berg laboratory had not set out with the specific objective of cloning genes, they had in fact made crucial observations in that regard. "Once Ron and I obtained these results it was clear to us that lots of folks, not just the ones in the Stanford biochemistry department, could readily make recombinant DNA," Mertz stated.[11] In correspondence with the author, Mertz wrote: *"I absolutely believe that I could have successfully cloned SV40 DNA into E. coli during the summer of 1972. With the recombinants already in hand it would have been rather trivial. But the self-imposed moratorium in the Berg laboratory precluded doing this* [author's italics]."[12]

When these experiments were completed, Berg, as a gesture of professional courtesy, instructed Mertz and Morrow to inform Herb Boyer (the source of the *Eco*RI enzyme) of their findings. "Herb came flying up here within the hour — and we went through the data," Berg stated. "It was decided that we would characterize the *Eco*R1 generated ends and he would sequence them." Boyer's paper, entitled "The DNA Nucleotide Sequence Restricted by the R1 Endonuclease"[13] was communicated to the *PNAS* by none other than Berg and was also published in the historic November 1972 issue of the journal, which ultimately contained four papers either authored or communicated by Berg.

This historic issue of the *PNAS* also included an article bearing the title "Enzymatic Oligomerization of Bacteriophage P22 DNA and of Linear Simian Virus 40 DNA,"[14] by the Italian molecular biologist Vittorio Sgaramella, who was then working in Lederberg's laboratory in the Stanford Department of Genetics. During that period, Sgaramella reported the ability of a DNA ligase expressed uniquely by bacteriophage

T4 to join phage P22 DNA molecules with flush (non-cohesive) ends. This extended similar observations reported earlier using synthetic DNA generated in Gobind Khorana's laboratory for the total synthesis of the yeast tRNA[Ala] gene *in vitro*.[15]

In retrospect, Sgaramella's observations might conceivably have provided an alternative or complementary strategy for generating recombinant DNA molecules. But he made no explicit mention of this possibility. In fact, when interviewed for the MIT Oral History in May 1975, Ron Davis offered a pessimistic opinion about the utility of phage T4 DNA ligase for generating joint molecules. "When I first heard about the possibility of joining DNA molecules from Sgaramella's earlier work with T4 ligase I started thinking about the possibility of joining DNA molecules *in vitro*," Davis stated. "The difficulty with that approach is that it's very inefficient — very inefficient! It can't be used to construct joint molecules. It requires very high concentrations of ligase for very long periods of incubation. And in the final analysis you get very few circular molecules. ·············· We realized ········ that the RI enzyme was a much easier thing to work with; all you had to do was cleave and reseal — very little enzymology."[16]

"So who first made the discovery that *Eco*RI generates complementary DNA ends that can be joined by [DNA] ligase?" Stanley Cohen (a Stanford professor of genetics whose role in the recombinant DNA story is more fully addressed in a later chapter) asked rhetorically in an interview by Sally Hughes in 2009. "I think that most people credit Mertz and Davis for that finding," he elaborated, "even though the papers by Mertz and Davis and by Vittorio Sgaramella were published in the same issue of the *PNAS*. The paper by Sgaramella (communicated to the *PNAS* by Lederberg) cites the findings by Mertz and Davis and acknowledges the use of biochemistry department facilities and expertise."[17] But, as just stated, it notably does not address the conclusions drawn by the pair of biochemists.

****

Another annual Progress Report to the American Cancer Society dated September 28, 1971 reported progress to date in the Berg laboratory:

> In the last report I outlined an approach for covalently inserting new segments of natural DNA into SV40 and Py DNA molecules. The purpose of this was to ·············· determine if bacterial genes could be integrated into animal cell genomes; and if so determine whether these bacterial genes could be transcribed and regulated. Much of the technology for performing this has now been worked out by Drs. David Jackson and Robert Symons. The following has been accomplished:
>
> (a) covalently closed circular SV40 DNA can be cleaved to produce full-length linear molecules.
> (b) using terminal transferase conditions have been developed for adding variable lengths (tails) of a single deoxynucleotide ··············.
> (c) synthetic polydeoxynucleotides dA:dT, dG:dC with short tails ··········· have been prepared.
> (d) hydrogen-bonded complexes between the synthetic polymers and modified SV40 linears have been prepared, but as yet covalently closed complexes have not yet been made.
> (e) the conditions whereby DNA polymerases and polynucleotide ligase can act together to fill in "gaps" and effect ring closure have been worked out with model compounds ···········.
> (f) a potential DNA insert, λ*dvgal* DNA (a hybrid construct carrying the entire *E. coli* galactose operon and a particular mutant form of bacteriophage λ called λ*dv* [see later discussion]) has been developed. As soon as the technology for DNA joining and ring closure has been solved, we expect to try to generate the suitable derivatized linears of λ*dvgal* in preparation for synthesizing the hybrid DNA molecule SV40-λ*dvgal*.[18]

Later in this report, Berg presciently wrote: "It seems fair to say that the possibilities opened up by this technique [referring of course to recombinant DNA technology] are quite novel."[19] However, he steadfastly maintains that he was then still not actively thinking about *cloning* individual pieces of DNA in the sense that we now use that term.

Notes

1. HB interview by Jane Gitschier, September 2009. Wonderful Life: An Interview with Herb Boyer. *PLoS Genet*, September 2009, e1000653.
2. *Ibid.*
3. *Ibid.*
4. *PNAS* **69**: 2904–2909, 1972.
5. *PNAS* **69**: 3365–3369,1972.
6. ECF interview with Janet Mertz, February 2012.
7. *Ibid.*
8. *Ibid.*
9. *Ibid.*
10. Mertz J, Berg P. (1972) Cleavage of DNA by R₁ Restriction Endonuclease Generates Cohesive Ends. *PNAS* **69**: 3370–3374.
11. ECF interview with Janet Mertz, February 2012.
12. E-mail from Janet Mertz, May 2, 2012.
13. Hedgpeth J, Goodman HM, Boyer HW. (1972) *PNAS* **69**: 3448–3452.
14. Sgaramella V. (1972) *PNAS* **69**: 3389–3393.
15. Sgaramella V, van de Sande JH, Khorana HG. (1970) *PNAS* **67**: 1468–1475.
16. RD interview by Rae Goodell, MIT Oral History Program, May, 1975.
17. SC interview by Sally Hughes, Regional Oral History Office of the Bancroft Library at the University of California, Berkeley, 2009.
18. Progress Report to American Cancer Society, September 18, 1971.
19. *Ibid.*

CHAPTER THIRTEEN

"Coincidence is the Word We Use When We Can't See the Levers and Pulleys"

(*Emma Bull, writer*)

As mentioned earlier, biochemistry graduate student Peter Lobban in Dale Kaiser's Stanford laboratory was independently conducting experiments startlingly similar to those being executed by the Berg group. The pair devised essentially identical protocols for generating recombinant DNA molecules in the test tube, at roughly the same time. Conspicuously, both protocols ultilized terminal transferase. Amazingly, a third research group headed by R.H. Jensen, R.J. Wodzinski and M.H. Rogoff also executed experiments using terminal transferase to join DNA molecules, though their 1971 paper in *Biochem Biophys Res Comm* entitled "Enzymatic Addition of Cohesive Ends to T7 DNA"[1] made no mention of recombinant DNA or gene cloning and the group never pursued these observations.

Coincidences of this type are typically considered to transpire by chance. But as detailed in a book entitled *Sociology of Science: Theoretical and Empirical Observations* published in 1979, sociologist Robert Merton examined the frequency of multiple simultaneous scientific findings. He identified a number of instances of multiple more or less simultaneous discoveries, including some with as many as six or eight people announcing the same finding.[2] He suggested that such "coincidental" events might transpire when timely scientific questions are "in the air" and multiple individuals are thinking about them at the same time — perhaps each from a slightly different perspective — rather than by pure coincidence. A dramatic contemporary example of this sort of hypothetical situation stems from the recent discovery of the Higgs boson. Five living theorists claim to having dreamed up this famous subatomic particle in physics.[3] Merton also suggested that the phenomenon may transpire more often than most of us think, since if one scientist makes a discovery in close temporal proximity to another, the later discoverer may forego announcing it publicly.[4]

....

Enter Peter Lobban, another Stanford biochemistry graduate student in the recombinant DNA story who completed undergraduate training at MIT. Like Janet Mertz, he majored in both the life sciences and engineering. Lobban's father was for many years a consultant chemical engineer, and Peter was sometimes torn between engineering and biochemistry as a career choice.

"When I talked with my advisor at MIT about graduate schools in biochemistry he told me that the best one in the country was at Stanford," Lobban related. "I applied to the program and visited Stanford for an interview one day in February when it was raining and sleeting in Boston — and sunny and warm in Palo Alto! So I was sold on the weather aspect of things right away!" Lobban was thrilled to meet the biochemistry faculty (who he described as luminaries) and departed Stanford excited at the prospect of studying with some of them in the near future.[5]

It's impossible to miss Lobban's patent intelligence, affirming the sincerity of Dale Kaiser's observation that "Peter is perhaps the brightest graduate student I have ever encountered."[6] Lobban projects a quiet sense of self-assurance and an unassuming demeanor strikingly reminiscent of Kaiser, his PhD mentor. "I wasn't aware of it at the time, but looking back I think that's true," Lobban commented. "I'm an independent cuss and Dale seemed like the sort of person who would be adaptable to that and wouldn't try to win a battle of wills with me. And that worked out very well."[7]

Lobban's research under Kaiser's mentorship was proceeding apace, though by his own admission: "I was at a stage where I really didn't know whether it would continue to go well or not."[8] Regardless, neither he nor Kaiser had any cause for concern in that regard, and switching research projects was then the furthest thing from Lobban's mind. As mentioned earlier, Kaiser was a noted authority on bacteriophage λ and Lobban's thesis research involved understanding the enzymatic mechanism by which λ integrated into the *E. coli* genome. In his own words:

> I started working on a project to discover and isolate the enzyme that was assumed to be necessary for the integration of the genome of λ bacteriophage into the *E. coli* genome. [The enzyme was discovered much later and is now known as "integrase"] ·············· For [this] purpose, I developed an assay system. It was assumed that integrase would mediate site-specific recombination between two pieces of λ DNA as well as between λ DNA and the DNA of *E. coli*. Therefore, the assay system I developed was one that would detect *in vitro* recombination between two different λ mutants, one with a pair of suppressor-sensitive mutations at one end of the genome and one with a pair of mutations at the other end, to generate a wild-type DNA molecule. ·····························[9]

Fulfilling the requirements for a Stanford doctoral degree in biochemistry required that students submit and successfully defend three original research proposals unrelated to ongoing research that might lead to a PhD thesis — and/or successfully endure three written examinations — or any combination of three research proposals and examinations. Lobban had decided to take his third (so-called qualifying) examination as a research proposal, so he took time away from the laboratory to prepare for this ordeal.

In the midst of this preparation, Lobban attended a biochemistry department journal club held on a regular basis, attended by faculty, post-doctoral fellows and graduate students. Berg has no recollection of being present at this journal club meeting at which the speaker was another biochemistry graduate student, Tom Broker, now a professor at the University of Alabama. Broker's seminar elicited the following response from Lobban:

> Tom presented a talk on terminal transferase. At the very end of his presentation, when some people were already starting to leave the room,

he threw out a statement to the effect that the enzyme could probably be used to put cohesive ends on DNA. I doubt that anyone other than me absorbed and retained that information. So when Broker mentioned using terminal transferase to make DNA with cohesive ends, a light bulb went on in my head and I said to myself, "Bingo! There's my third examination topic!"

The proposal I wrote involved developing a method for inserting DNA from any desired source into the λ genome *using terminal transferase to grow complementary cohesive ends on the two DNA molecules that were to be joined* [author's italics]. When the DNAs were mixed, it was expected that the complementary cohesive ends would anneal to one another and then DNA polymerase and DNA ligase would be used to fill in any gaps that might be present and make the joining covalent. The resulting hybrid DNA molecule would then be introduced into *E. coli* by ·········· [transduction] and an appropriate selection method would be used to select for a λ bacteriophage variant containing the DNA of interest from the original source.[10]

The epiphany that visited Lobban at the end of Broker's seminar was the theoretical potential of using terminal transferse to join two different DNAs that possessed complementary cohesive ends, rather than by recombination in living cells. When asked by the author why this was a "Bingo moment" for him, Lobban replied:

At that point I was steeped in the uses of *in vivo* recombinant DNA to do genetic probing, and also to construct organisms that would make something (proteins). So I figured that anything that would allow one to do that in a more general way would open up a whole field of tools and uses.

Just after I finished presenting my examination proposal, as Dale Kaiser and I were exiting the conference room, he asked me whether I wanted to change my dissertation project to the one I had just proposed. I said I did. In fact, though, what I did for my dissertation was to develop a model system for *in vitro* recombination that did not actually produce any useful recombinants but served as a direct and easy way to prove the concept, starting from a DNA that was easily obtained and had well-studied characteristics, namely, that of bacteriophage P22. The idea was to get the system for recombination working first and then to make useful recombinants using the method later. I ended up doing the first part of the job, but never got to the second part.[11]

Lobban's examination proposition was rooted in observations, including some from the Kaiser laboratory mentioned earlier, that bacteriophage λ DNA is naturally endowed with cohesive (sticky) ends. In fact, by his own admission, Berg's idea of opening viral DNA rings and preparing their ends for receiving a new DNA segment also emerged from his awareness of Kaiser's earlier studies on cohesive ends in linear λ DNA.

Peter Lobban. (Courtesy of Peter Lobban.)

In his examination proposal Lobban presciently wrote:

[*an*] *eventual goal for the method, assuming its success at simpler tasks, would be to produce a collection of transductants synthesizing the products of genes of higher organisms* [author's italics] This would be of immense help in the purification of the proteins made by these genes, for in many cases they would be far less dilute in the bacterium than in the cells of their origin. ⋯⋯⋯⋯. It may be fairly stated that there would be many uses for the transducing phage once they were made.[12]

Lobban did not use the term "gene cloning" in his third written examination. However, in late 1969, he was evidently thinking ahead (albeit loosely) to a time when individual genes might be expressed outside their normal environment to yield proteins of particular interest.

Berg was not on Lobban's third examination committee, a fact unequivocally confirmed by Lobban himself — and one not widely appreciated in the scientific community, perhaps because of the proximity of the examination and the subsequent defense of his PhD thesis to a committee that *did* include Berg. In fact, though Berg and his students became fully aware of Lobban's interest in generating recombinant DNA in the early 1970s and in fact exchanged pertinent technical information with him, Berg did not learn about the written or oral content of Lobban's third examination proposal until some years later.

"I can't recall specifically when I heard about Peter's proposal, but more than likely it emerged during one of our group meetings where we would have been discussing Dave's or Janet's progress," Berg related. "Long after the work had been completed and even published, Arthur Kornberg and I were discussing a chapter in his impending book *For the Love of Enzymes* that discussed Peter's idea and work. I told him I had not been at Peter's proposal meeting where he described his ideas nor had I ever seen Peter's written proposal. He then sent me the copy."[13]

An e-mail from Berg to Kaiser sometime in 2006 is pertinent in this regard:

Dale; ...., I was one of the signers of Peter's thesis but that was of the order of two years after Dave Jackson and Bob Symons had begun our experiments. However, to the best of my knowledge and recollection I was not on his Proposition Committee when Peter presented his ideas on recombination. It was only years later that I learned the full extent of what he had conceived when Arthur [Kornberg] gave me a copy of Peter's [examination] proposition.[14]

The belief by a substantial number of individuals interviewed (including Dale Kaiser when interviewed by the author in 2011) that Berg was a member of Lobban's examination committee appears to have led to the rash conclusion by some that Berg might have co-opted the idea of using terminal transferase for generating artificial cohesive ends in DNA from Lobban's presentation, a notion categorically denied by Berg. This mistaken belief likely arose (and stuck) in the minds of some who were confused about the fact that Berg was indeed a member of the PhD thesis committee that examined Lobban following the customary public defense of his thesis in May 1972, but not on his preceding third examination committee. Kaiser was apparently not educated to this confusion until he received the e-mail from Berg in mid-2006 stating otherwise.

When communicating with Berg and Kaiser in the course of my writing, neither individual could recall what prompted the electronic communication from Berg in 2006 reproduced above. The two faculty members were good friends and talked reasonably frequently. So it is entirely possible that Berg wished to correct or amend statements about past events that arose in conversation in or shortly before 2006. Conceivably such a conversation was prompted by Berg's reading of an autobiographical article by Kaiser published in the *Annual Reviews of Microbiology* in that year, in which Kaiser stated: "······ in 1971, as a graduate student, Peter presented a thesis proposal to his advisory committee, *which included Paul Berg* [author's italics] and Bob Lehman, describing the feasibility of joining two arbitrary DNA molecules using the purified enzymes he knew to be present in the freezers of Kornberg and Lehman down the hall."[15] Lobban relates that shortly after he (with Kaiser's support) elected to switch his PhD thesis research, "Dale casually asked [him]: 'Do you know that Paul Berg is also working on a similar idea?'"[16]

A review article by Berg and Janet Mertz published in 2010, bearing the title "Personal Reflections on the Origins and Emergence of Recombinant DNA Technology," provides their summation of the events related above, including a timeline of the emergence of the technology.[17]

Regarding the thankfully rarely encountered propensity to "steal" other scientists' work in progress, it is to be noted that familiarity with most, if not all ongoing research in the Stanford Department of Biochemistry, including those in training, was the rule rather than the exception under Arthur Kornberg's chairmanship. Kornberg steadfastly insisted on open scientific interaction between all faculty, postdoctoral fellows and graduate students in the department. To be sure, it was usual for graduate students and postdoctoral fellows with different mentors to work side by side in a laboratory independent of that "belonging" to either mentor.

"This was an aspect of the department of biochemistry that was both wonderful and problematic for me," Lobban related. "Arthur insisted that everyone in the department work together. You worked together; you shared reagents together; you talked to one another about science. We were all on a first name basis — except for referring to Arthur as Dr. Kornberg! So the notion of competition within the department did not exist. This provided a virtual intellectual paradise for working in. But it also gave one a distorted view of how the rest of the world functioned. I personally tend be a little naive and I tend to be more trusting than most. But my trust was not misplaced back then.[18]

"It may not have been immediately after my third examination in November 1969 that I met with Paul," Lobban continued. "But soon after that either he or Dave Jackson told me that they were going to acquire terminal transferase and that they would happily give me some of that enzyme if I wanted it."[19] "And when I purified λ exonuclease* I gave some of that enzyme to them. Using that enzyme was not an obligatory part of their experimental protocol. I was simply able to show that if one did not use λ exonuclease the terminal transferase reaction had a lag phase consistent with the notion that the enzyme preferred single-stranded DNA primers, but could work on double-stranded DNA. I didn't take the time to prove this experimentally, but I told them that after using λ exonuclease the transferase reaction was linear from the outset. And so they used it."[20]

The use of λ exonuclease provided a significant lift for Jackson and his collaborator Bob Symons, who were grappling with the inefficiency of generating cohesive ends with terminal transferase. "I discovered that the terminal transferase was not initiating at all well at

---

* Exonucleases are enzymes that degrade DNA from an end of one strand (including ends generated by single stranded nicks or gaps with relevant 3' or 5' termini). Lambda exonuclease is one of a number of such enzymes, and proved to be extremely useful for improving the efficiency of adding individual nucleotides by terminal transferase by lengthening the size of single-stranded ends prior to adding the transferase. [Interview by Rae Goodell, MIT Oral History Program, March 1978].

the ends of linear SV40 DNA," Jackson stated in a 1978 interview. "Only a few percent of the molecules had any tails. It was Peter Lobban who came up with a solution to that, because he was having the same trouble with P22 DNA. He thought that the problem was that the enzyme might not initiate well on blunt-ended molecules and that it might be necessary to have the strand on which the terminal transferase initiates sticking out as a short piece of single-stranded DNA. So he suggested using λ exonuclease. It was tricky to remove just a few nucleotides from a long DNA molecule — and do so for every DNA molecule. But once we solved this problem terminal transferase did exactly what we wanted."[21]

Berg was similarly effusive in his praise of Lobban's contributions: "Peter — and we refer to him in our November 1972 *PNAS* paper — made several critical discoveries that facilitated the technique. For example, when you cut DNA and both ends are either flush or very close to being flush it's very difficult for terminal transferase to add nucleotides onto the end. On the other hand, if one peels back the 5-prime strand, causing the 3-prime end to stick out, the enzyme adds nucleotides very well. Peter discovered that. He very quickly communicated that to us, indicating that if we were going to make cohesive ends it was useful to pre-treat the DNA with an enzyme that cuts back the 5-prime end a little. We acknowledged Peter's contribution in our paper. ...."[22]

"There was close communication between the two labs," Berg continued. "But it wasn't a formal collaboration. We never intended to work on the same project. *Since I'd never seen his third examination proposition I was unaware of the implications he foresaw for being able to do this* [author's italics]. But as it turned out we were trying to recombine SV40 DNA and a DNA molecule called λ*dvgal*, while Peter was trying to join two phage P22 DNA linear molecules. So I suggested to that we get together with Peter and see whether it might be helpful to work in parallel — or in concert. Our research was strictly motivated by trying to get bacterial DNA attached to SV40 so that we could get it into mammalian cells."[23] Lobban concurred with this opinion. "We agreed to work in parallel and to share information and reagents," he stated.[24]

A letter of recommendation from Berg written on May 22, 1973 was unabashedly complimentary of Lobban:

> Peter worked with Dale Kaiser but I got to know him well because the problem he worked on for his PhD thesis was closely related to work we were doing. It's of interest that he developed his PhD thesis as a proposition for his preliminary examination. It was so novel in its conception and beautifully thought through that he was encouraged to drop the work he was doing to concentrate on developing the techniques for joining DNA molecules together. I and Dave Jackson had independently hit on the same idea for inserting DNA into SV40; at all times we had a very amicable, cooperative and most important "collaboration", so much so, that it was Peter who often made the key breakthrough to solving technical problems.[24A]

Clearly influenced by his engineering orientation when solving biochemical problems, after completing his third examination and returning to the bench, Lobban refrained from immediately delving into tailoring phage λ DNA. As mentioned above, he elected to develop a protocol for generating dimers of phage P22 DNA from P22 DNA monomers as a proof of principle before attempting to construct a λ transducing phage. His reasoning was that phage P22 was a more appropriate model system for working out the detailed methodology for linearizing, joining and recircularizing λ DNA, since the P22 genome is both linear and blunt-ended, thus precluding dimerization of DNA molecules without the addition of (poly)dAMP and (poly)dTMP tails. His studies with P22 DNA adequately fulfilled the requirements of his PhD thesis — and Lobban never got around to working with phage λ as a graduate student.

There is no paucity of documented evidence in support of the open and cordial communication between the Jackson/Symons/Berg and Lobban/Kaiser groups. The acknowledgement section of the 1972 paper in the *PNAS* by the former three individuals reads: "We are grateful to Peter Lobban for many helpful discussions." Correspondingly, Peter Lobban's PhD thesis offered "special thanks to Dr. David Jackson, who worked on a similar system while he was here and who shared ideas and results with me."[25] And when Lobban ultimately published the results of his work with Dale Kaiser in a lengthy article in the *Journal of Molecular Biology* in 1973, the authors acknowledged "the fruitful discussions and free exchange of ideas we enjoyed with Drs. David Jackson, Robert Symons and Paul Berg, who worked concurrently on a similar project."[26]

"Paul made no secret of talking about what we were doing and trying to do," David Jackson commented. "He is someone whose style of science is to talk to lots of people and tell them what he's doing, and listen to them tell what they're doing — and to be very open about it all. So anybody who cared to listen knew what we were trying to do."[27] Other former members of the Berg laboratory emphatically reinforced this sentiment.

Further support for the notion that the Berg and Kaiser laboratories independently hit on essentially the same idea at roughly the same time derives from a document (unpublished, and to the author's knowledge undistributed) that Kaiser drafted in late August 1980 entitled "Description of Work Leading up to Recombinant DNA." Kaiser does not recall what prompted him to write this piece. But he agreed with the author's suggestion that it was conceivably prepared in anticipation of remarks he might have to deliver (but ultimately did not) at impending ceremonies celebrating the 1980 Lasker Award in November of that year, of which he was a recipient *"for his crucial role in creating recombinant DNA methodology through his path breaking studies of cohesive single-stranded DNA."*[28] Notably, Kaiser's unpublished essay reads:

> This brings us to 1969 when Peter Lobban, a graduate student working with me, brought together three lines of work: infecting *E. coli* with lambda DNA, breaking DNA into pieces, and joining DNA pieces with sticky ends. ···.
>
> Lobban then set about demonstrating that his scheme for joining pieces of DNA molecules together really worked (Lobban and Kaiser, 1973). *During his tests of the feasibility of forming recombinant DNA molecules, Lobban collaborated with Jackson, Symons and Berg who, independently working in the same department, were constructing recombinant DNA molecules joining Simian Virus 40 and lambda DNA* [author's italics].[29]

The author's emphasis on the cordial and collegial attitude between the Kaiser and Berg laboratories during the period under scrutiny is significantly motivated by the fact that several published accounts of the emergence of recombinant DNA technology offer the view that Peter Lobban was not accorded the public recognition he deserved. In his book *Recombinant DNA — The Untold Story*, John Lear wrote:

> Lobban was named five times in the Jackson–Symonds–Berg report to *PNAS,* thanked "for many helpful discussions," and specifically credited with prior demonstration of the necessity for using certain enzymes that proved crucial to the success of the Berg team's experiment. It is therefore curious that Lobban should be almost totally neglected, as he has been, in popular accounts of the origin and development of one of the most pivotal, if not the most pivotal, series of scientific discoveries in the history of Homo sapiens.[30]

And when recounting some of these events in his 1989 autobiography, *For the Love of Enzymes — The Odyssey of a Biochemist*, Arthur Kornberg penned: "among the pioneers of recombinant DNA, Peter Lobban is not only unsung, but is no longer even in the chorus of basic science."[31]

The viewpoint that Lobban did not achieve the enduring and successful career as a biochemist he so richly deserved cannot be denied. But several explanations are tenable for this ill-fated outcome other than the interpretation that he was ignored for ulterior motives. First and perhaps foremost, when Berg, Jackson and Symonds were preparing their groundbreaking paper published in the *PNAS* in November 1972, Berg explicitly suggested to both Kaiser and Lobban that the two laboratories publish consecutive papers in the journal. He also offered to hold up the contribution from his own laboratory if a manuscript by Lobban and Kaiser was in the offing in the near future. Kaiser declined the offer. Some written accounts promulgate the view that when Berg discussed this notion with him, Kaiser was of the opinion that Lobban's studies were not ready for publication. "It seemed reasonable that brief reports of the success of both groups should appear together in an early issue of the *Proceedings of the National Academy of Sciences*," Kornberg wrote in his autobiography. "Kaiser preferred, however, that some loose ends in Lobban's thesis work be completed."[32]

"My own desire to publish our findings when we did was significantly motivated by the fact that Dave Jackson was leaving the lab," Berg related. "He had accepted a position at the University of Michigan and wanted to leave before even writing that paper. But I told him: 'No way! If you leave before we've written this it will end up dragging out forever.' I actually went so far as to write to the individual who recruited David to the University of Michigan, imploring him to extend Peter's start date. But when I talked to Dale he related that he and Peter would not be able to meet the same time schedule as us and that he wanted their paper to be more detailed than what might be accommodated in the limited size of articles allowed in the *PNAS*."[33]

Lobban also offered a version of the events that transpired after Berg's offer. "I specifically remember mentioning Paul's offer to Dale," he stated. "I have a vivid mental picture of this happening right outside my lab — perhaps in the corridor. He did not say anything to me of competition within the department. Both of us knew that a paper in the *PNAS* would be a preliminary one, as was the Jackson, Symons and Berg paper. But my reaction was necessarily: 'He's my advisor and I have to respect his opinions.' I'm told that Dale now regrets this decision."[34]

Regardless of this unfortunate outcome Lobban was privately unambiguously grateful for the opportunity to publish alongside the Berg group, despite the fact that this suggestion never materialized. In early July 1972, he wrote a warm letter to Berg informing him of his progress in Canada, where he was pursuing postdoctoral training with Louis Siminovitch. He used that opportunity to thank Berg for his efforts:

> Greetings from Toronto! Settling down here took a long time, but I am finally in the lab full-time and starting work. David Jackson called today to discuss the paper you and he have written for *PNAS* and to ask about our publication plans. Basically, the situation is that Dale is opposed to our writing a *PNAS* paper, so we will only do the *JMB* paper. Thus you and Dave should feel free to submit your paper as soon as it is ready.
>
>   When we discussed the idea that both groups would publish back-to-back in *PNAS* I was not aware how reluctant Dale would be to go

along with that idea. So I beg pardon for any confusion that may have been caused. I appreciate greatly that you and Dave checked with me before publishing, and that you were willing to wait for us to complete a paper.[35]

Berg does not recall this clarification from Lobban buried in his Stanford University archives. "It's interesting that Dale might have worried about the perception of internal competition," he commented when shown Lobban's letter. "Knowing him, I believe he probably had that instinct, but didn't realize perhaps that publishing together would have had exactly the opposite effect; it would have eliminated any thoughts in anyone's mind about competition."[36]

Lobban asked Berg to critically comment on the paper by him and Kaiser then in preparation for the *Journal of Molecular Biology*. Berg did so. On December 18, 1972 Lobban again wrote to Berg, stating: "Here is the second draft of our paper. As you can see, I took your advice and pruned it extensively.[37] To which Berg graciously replied:

Dear Peter, I think the paper is elegant! Much improved over the earlier version. ⋯⋯⋯⋯⋯⋯⋯⋯⋯⋯⋯⋯⋯. Everything in the department is going well and we certainly miss you around here. Come back and visit when you can. You're always welcome."[38]

When Lobban's work was eventually published in the *Journal of Molecular Biology* in August 1973 it occupied fully 18 pages of that revered journal — a lengthy article by any measure. But the concluding paragraph offered only a cursory consideration of "the apparent generality of the method described here for the joining of any pair of double-stranded DNA molecules to each other ⋯⋯⋯."[39] And the phrase "recombinant DNA technology" is notably absent. In contrast, the concluding sentences in the introductory paragraph of the paper by Jackson, Symons and Berg published in October 1972 unambiguously trumpeted the exciting potential of their studies.

⋯⋯ we have developed biochemical techniques that are generally applicable for covalently joining any two DNA molecules. Using these techniques, we have constructed circular dimers of SV40 DNA; more-over, a DNA segment containing λ phage genes and the galactose operon of *Escherichia coli* has been covalently integrated into the circular SV40 DNA molecule. Such hybrid DNA molecules and others like them can be tested for their capacity to transduce foreign DNA sequences into mammalian cells, and can be used to determine whether these new non-viral genes can be expressed in a novel environment.[40]

This being said, there is little question that Lobban was indeed harboring a vision of what we now define as gene cloning. As pointed out by historian Doogab Yi:

> Lobban foresaw the potential use of a recombinant DNA transduction system for "genetic engineering" in both his PhD proposal and dissertation. Lobban's suggestion for expressing recombinant DNAs in a foreign host came from his initial interest in problems of antibody synthesis: recombinant DNA-mediated transduction could provide a source for new antibodies if transduced genes could be expressed in the host. In his 1969 [examination] proposal he indeed speculated on a potential use of his enzymatic DNA-joining method: if transduced hybrid DNAs could directly express their new genetic information, re-combinant DNA-mediated transduction could provide a new source for useful gene products. Based on the possibility of recombinant-mediated transformation, Lobban suggested that the ability to insert and express recombinant DNA molecules in bacteria could become a new technique in biomedical research. Bacteria could become "factories" to produce medically useful gene products like antibodies. As he put it in his proposal and dissertation, "if the bacterial host of the transducing genomes [bearing mammalian genes] is able to transcribe and translate them, it could be used as a source of the gene product that might be far more convenient than the mammalian cells themselves."[41]

Such considerations were in fact uppermost in Lobban's thoughts when, after realizing his PhD degree, he went to Canada for postdoctoral training. He identified Simonovitch's laboratory in Toronto as an environment in which he might acquire the skill sets necessary to pursue work with mammalian cells. Had he considered postdoctoral training close to home with someone like Berg, or Renato Dulbecco, things might have turned out very differently for him. While in Toronto, Lobban applied widely for an academic position in the United States. He was disappointingly unsuccessful, despite a glowing recommendation from Berg, who wrote on his behalf:

> Having followed his work closely I can say that he is one of the most careful and critical young people we've ever had. Moreover, he is skillful technically and exhibits no inertial barrier to tackling new things. Coupling this with a highly developed creativity, I would predict a very promising future for Peter. He works hard and he has a very strong commitment to science. All in all I think he's a very fine candidate and I recommend him to you unequivocally.[42]

Another explanation for the unfortunate manner in which Lobban's scientific career unfolded derives from events that transpired after he completed his postdoctoral training. For one thing, his search for an academic position was unfortunately mistimed. Financial support for basic science was then tight and junior faculty positions were hard to come by. Additionally and perhaps more significantly, his prescience about the potential of recombinant technology in his third examination proposal and in his PhD thesis (neither of which were then readily available outside Stanford University) was too fresh to be widely appreciated. Indeed, few if any members of the scientific community recognized, or were then even aware, of its phenomenal promise. After listening to Lobban describe his future research intentions, a faculty member at one of the institutions to which Lobban was invited to interview for a faculty position dismissively told him that his experimental strategies had no chance of working! Then too, when visiting more than a dozen academic institutions to which he was invited to interview, Lobban inexplicably elected to mainly present seminars on his postdoctoral work in Toronto (which though solid, was not in any way ground-breaking) rather than his Ph.D thesis work.

Complicating this strategic miscalculation, Lobban is described by many as a somewhat reclusive and "laid-back" personality who may not have projected a particularly engaging persona on interview. Kornberg referred to him as "diffident and shy,"[43] and Berg commented on his "unambitious personality"[44] — sentiments that Lobban does not dispute.

Having relinquished hope of establishing a career in biochemistry/molecular biology, Lobban returned to Stanford to take a master's degree in electrical engineering and went on to enjoy a highly successful and gratifying career in the private sector. He offers no obvious hints of bitterness about Kaiser's decision not to publish together with Berg, *et al.* "I'm not a terribly ambitious or competitive person. Dale was my professor and I was obliged to do as he asked."[45]

"By the time I graduated from the Stanford biochemistry program *Genentech* (a biotechnology company in San Francisco) was starting up," Lobban continued. "I've often thought that had I joined that company instead of going to Siminovitch's lab I might have at least made money," he stated sardonically, "even though I might not have been happier than I am now." A palpable silence followed before Lobban continued with an ironic smile. "Still, I wouldn't be human if I didn't ask myself: 'Well, if we'd published back-to-back — might I have shared a Nobel prize?'"[46]

"I personally believe that at that time nobody, except perhaps Peter — who stated it in his examination proposition and in his thesis — believed that this sort of work was going to make a big splash," Berg commented. "The reality is that it did! More than I ever anticipated."[47] But in the final analysis Berg is no more willing to categorically credit Lobban with the explicit idea of gene cloning than he is to credit anyone else at that time — including himself. "He didn't advance the idea of gene cloning," Berg told interviewer Sally Hughes. "He advanced the notion that constructing molecules that could be introduced into mammalian cells might permit expression of the introduced DNA segment and perhaps the isolation of significant amounts of proteins encoded by genes contained therein."[48]

Notes

1. *Biochem Biophys Res Comm* **43**: 384–392 (1971).
2. http://www.quora.com/Knowledge-Discovery/Why-does-multiple-discovery-happen
3. Cho A. (2012) *SCIENCE* **337**: 1286–1289.

4. http://www.quora.com/Knowledge-Discovery/Why-does-multiple-discovery-happen

5. ECF interview with Peter Lobban, September 2011.

6. ECF interview with Dale Kaiser, April, 2011.

7. ECF interview with Peter Lobban, September 2011.

8. *Ibid.*

9. Telephone interview with Peter Lobban, May, 2013.

10. ECF interview with Peter Lobban, September 2011.

11. *Ibid.*

12. Lobban P. Third exam proposal, Stanford Department of Biochemistry, November 6, 1969.

13. Interview with Paul Berg, June 2012.

14. E-mail from Berg to Kaiser dated ⋯⋯ . Courtesy of Dale Kaiser.

15. Kaiser DA. (2006) A microbial genetic journey. *Annu Rev Microbiol* **60**: 1–25.

16. ECF interview with Peter Lobban, September 2011.

17. Berg P, Mertz J. (2011) Personal Reflections on the Origin and Emergence of Recombinant DNA Technology. *Genetics* **184**: 9–17.

18. ECF interview with Peter Lobban, September 2011.

19. *Ibid.*

20. *Ibid.*

21. Interview with Rae Goodell, MIT Oral History Project, 1978.

22. Interview with Paul Berg, June 2012.

23. *Ibid.*

24. ECF interview with Peter Lobban, September 2011.

24A. Letter from Paul Berg to Hilary Koprowski, May 22, 1973.

25. Jackson D A, Symons RH, Berg P. (1972) Biochemical method for inserting new genetic information into DNA of Simian Virus 40: circular SV40 DNA molecules containing lambda phage genes and the galactose operon of *Escherichia coli*. *PNAS* **69**: 2904–2909.

26. Lobban PE, Kaiser AD. (1973) *J Mol Biol* **78**: 453–471.

27. ECF phone interview with David Jackson, April 2012.

28. http://www.laskerfoundation.org/awards/1980_b_description.htm

29. Dale Kaiser, personal communication, September 2011.

30. Lear J. (1978) Recombinant DNA: The Untold Story. Crown Publishers Inc., p. 39.

31. Kornberg A. (1989) For the Love of Enzymes: The Odyssey of a Biochemist. Harvard University Press, p. 280.
32. *Ibid.*
33. ECF interview with Paul Berg, June 2012.
34. Interview with Peter Lobban, September 2011.
35. Letter from Peter Lobban to Paul Berg, 5 July 1972. Reproduced with permission from the Department of Special Collections, Stanford University Libraries.
36. Interview with Paul Berg, June 2012.
37. Letter from Peter Lobban to Paul Berg, December 18, 1972. Reproduced with permission from the Department of Special Collections, Stanford University Libraries.
38. Letter from Paul Berg to Peter Lobban, ········ 1972. Reproduced with permission from the Department of Special Collections, Stanford University Libraries.
39. Lobban PE, Kaiser DA. (1973) Enzymatic end-to-end joining of DNA molecules. *J Mol Biol* **78**: 453–460.
40. Jackson DA, Symons RH, Berg P. (1972) Biochemical method for inserting new genetic information into DNA of Simian Virus 40: circular SV40 DNA molecules containing lambda phage genes and the galactose operon of Escherichia coli. *PNAS* **69**: 2904–2909.
41. Doogab Yi. ········
42. Letter from Paul Berg to various individuals, 1973–1974. Reproduced with permission from the Department of Special Collections, Stanford University Libraries.
43. Kornberg A. (1989) For the Love of Enzymes: The Odyssey of a Biochemist. Harvard University Press, p. 280.
44. Interview with Paul Berg, September, 2011.
45. ECF interview with Peter Lobban, September 2011.
46. *Ibid.*
47. Interview with Paul Berg, September, 2011.
48. Interview by Sally Hughes, Regional Oral History Office of the Bancroft Library at the University of California, Berkeley, 1997.

CHAPTER FOURTEEN

Yet Another Stanford Contribution

*I was confident that Paul Berg would support
an offer of a faculty position at Stanford*

Bespectacled, balding and bearded, New York-born Stanley (Stan) Cohen proffers the physical persona of a Talmudic scholar. Such superficial evaluations certainly don't conjure up images of a 1950s "pop" music writer and performer. Cohen's flair for pop music (as a teenager he played the piano, ukulele and guitar) inspired him to organize a band while a college student at Rutgers University — and to write several original songs under the pseudonym Norman Stanton, a play on his first and middle names — Stanley Norman.

The celebrated American vocalist Billy Eckstine recorded one of Cohen's compositions, entitled *Only You*. "The song began to take off on the hit parade," Cohen proudly told interviewer Sally Hughes,[1] "But its popularity was smothered by another 1950s song in the sentimental genre called *Only You and You Alone*, written by another composer, Buck Ram, who wrote, produced and/or arranged music for performing greats that included The Platters, The Drifters, Ike & Tina Turner, Glenn Miller, Duke Ellington and Ella Fitzgerald. *Only You and You Alone* was eventually shortened to *Only You* and was recorded by *The Platters* in 1955. That rendition (a nostalgic favorite to this day) leaped to the top of the charts!" Nonetheless, royalties that Cohen received for his musical efforts helped offset some of his college expenses — and he took considerable pleasure from that time. In his mid-70s at the time of this writing, Cohen not infrequently accompanies his daughter Anne, a talented banjoist in her own right.

Despite entreaties to launch a career in the music publishing business, Cohen came to the commonsensical conclusion that his intellectual skills far exceeded his musical panache! So in 1957 he entered medical school at the University of Pennsylvania. Following graduation he completed his internship and residency training in internal medicine and harbored aspirations of pursuing biomedical research in an academic environment — preferably a research-oriented department of medicine. But unanticipated events dictated otherwise. During his final year at medical school, the Vietnam conflict was raging and Cohen was called to military service as a physician. Scornful of the prospect of providing standard health care to non-combat military personnel, he applied for and received one of the coveted commissions available to a fortunate few research-minded physicians who served out their military obligation at the National Institutes of Health, in Bethesda, Maryland. The United States government may be gratified to know that during several contemporary wars, a significant cadre of such military "escapees" brought far more distinction to the country than they would have had they served the war effort diagnosing hernias!

When his two-year military obligation drew to a close, Cohen decided to expand his biochemical and molecular biological horizons. He secured a postdoctoral position with Berg's long-standing friend and scientific colleague from graduate student days, Jerry Hurwitz. Then a faculty member with his own research program at the Albert Einstein College of Medicine in New York, Hurwitz's laboratory was not infrequently visited by colleagues from other institutions. During his postdoctoral sojourn, Cohen met the Stanford biochemistry stalwarts Berg, Kornberg and Kaiser.

Under Hurwitz's watchful eye, Cohen's contributions to the thriving field of phage and bacterial genetics earned him notice in the bacteriophage λ community. Not all the notice was overtly complimentary, from his point of view. "Mark Ptashne once invited me to give a talk at Harvard when [Jim] Watson was the department chair," Cohen related. Unaware that reading newspapers during research seminars was one of Watson's favorite idiosyncrasies,* Cohen became somewhat anxious. After all, presenting his work at Harvard as a mere postdoctoral fellow, in the presence of Jim Watson no less, was a major event in his career. Not surprisingly, he was disturbed that Watson seemed to find his work boring, as evidenced by his engagement with his newspaper while the seminar was in progress. "But at the end of the seminar he asked a number of insightful questions," Cohen recounted, "So he'd apparently been listening,"[2] (as Watson typically does when a seminar topic interests him, notwithstanding the omnipresent newspaper on his lap).

Cohen took full advantage of the geographical proximity of the Albert Einstein College of Medicine to the Cold Spring Harbor Laboratory on nearby Long Island. He sat in on several summer courses there and was especially attentive to a lecture by Richard Novick from New York University, who was then working on plasmids that conferred antibiotic resistance to Staphylococci. Cohen's postdoctoral work yielded several notable publications in well-regarded journals and Hurwitz was sufficiently impressed that he encouraged the young postdoctoral fellow to seek a faculty position. "But I wasn't sufficiently confident that I could make it in the arena of phage lamdology," Cohen related. "Besides, I had been trained as a physician and I enjoyed teaching medicine."[3] Cohen consequently sought a position in an academic medicine department — one in which he could usefully meld his mounting expertise as a basic scientist to his clinical expertise. In due course, he was offered a junior faculty appointment in the Department of Medicine at the University of Pennsylvania. However, shortly before he was due to initiate his career as a physician–scientist, the chair of the department, James (Jim) Wyngarten, relocated to Duke University. "He wanted me to join him there," Cohen commented,[4] "But I had worked as a resident at Duke and wasn't keen on

---

* Watson's penchant for reading newspapers in the midst of scientific presentations was one of his well-known idiosyncrasies. At a scientific conference in Copenhagen sometime in the 1960s, the ever-mischievous Sydney Brenner covertly distributed newspapers to the assembled conferees (except Watson) with instruction that they should feign reading them when it was Watson's turn to speak!

living in North Carolina." (It's noteworthy how discriminatory young scientists such as Berg and Cohen could afford to be about places to live in those early days of molecular biology!)

Cohen's decision not to compete directly at the forefront of the basic science arena was reinforced by his observation that the burgeoning field of phage λ genetics was highly competitive — "too competitive for my taste," he declared. (Cohen once calculated in jest that if one divided the number of nucleotides in the human genome by the number of recognized phage lambdologists, there was about 1 per 50 nucleotides — "some just a few nucleotides apart!"[5] So he decided to focus on plasmid biology, a field strongly influenced by work on bacterial antibiotic resistance. Indeed, when debating whether or not to join the faculty at the University of Pennsylvania, Cohen frequently visited microbiologist Stanley (Stan) Falkow, then at Georgetown University, to procure bacterial plasmids that conferred antibiotic resistance. In anticipation of a move to the University of Pennsylvania Cohen also submitted a research proposal on this topic to the NIH.

Hurwitz shared his laudatory opinion of Cohen with his Stanford colleagues, one or more of whom likely passed this information on to Halsted (Hal) Holman, chair of the Stanford Department of Medicine. In due course, Cohen was invited to present seminars in both the Departments of Medicine and Biochemistry. "My seminar in the medicine department was poorly attended," he related. "But the biochemistry library (where research seminars were routinely held) was packed and it was clear that a lot of faculty, students and postdocs were interested in the work I was doing in Hurwitz's lab. There were good questions and enthusiastic discussion. It was apparent to me that I had passed muster — and I was confident that Paul Berg would support an offer of a faculty position at Stanford."[6]

Berg was indeed an enthusiastic advocate. Years later he wrote a letter to Holman in support of Cohen's promotion to tenure in the Department of Medicine at Stanford:

> I first met Stan in Jerry Hurwitz's lab. Hurwitz was very impressed with his intelligence, his drive and his determination and effectiveness at

the lab bench. Knowing Hurwitz to be very critical and to have high standards I accepted that evaluation and as you'll recall, supported his appointment when he came to Stanford.[7]

In this letter, Berg also specifically commented favorably on Cohen's work with plasmids that conferred antibiotic resistance.

> I [am] enthusiastic about Stan's work on the R-factor mediated transfer of drug resistance genes. Here he has made, I believe ···· meaningful and original contributions. ········· From my reading and listening to the R-factor work I'm impressed with the sophistication of his understanding and technical skills. Because he knows the molecular biology and genetics of plasmids he is able to provide a very professional approach to a study of the R-transfer factors, how they pick up and transfer drug-resistance genes ····[8]

Berg commented too on aspects of Cohen's sometimes anxious personality:

> Having watched and talked to Stan many times since he arrived, I'm impressed by his drive. Very likely it's that drive which creates problems in his interactions with others. Nevertheless if that does not neutralize his effectiveness or does not create problems with his immediate colleagues then I would say that the internal drive will probably ensure his scientific productivity for many years to come.[9]

Stanley N. Cohen. (www.nature.com/nrm/journal/v9/n8/full/nrm2458.html)

During his visit to Stanford, Cohen met with the venerable Arthur Kornberg. But he was taken aback by Kornberg's skepticism about simultaneously managing vocations in both clinical medicine and basic science. "Regardless of the obvious relationship between antibiotic resistance and plasmid biology Arthur was also dubious about the value of studying plasmids in general," Cohen related. "He told me that most of the things that were important to know about plasmids were probably already known — and that I should study something meaningful — like bacteriophage!"[10]

Hal Holman on the other hand was impressed with Cohen's credentials as a research-oriented physician and offered Cohen a junior faculty position in the Department of Medicine, with the understanding that in addition to conducting basic research, Cohen would assume clinical responsibilities in the division of hematology. This arrangement suited Cohen, who was not at all averse to honing some of his clinical skills. Meanwhile, his research grant proposal on plasmid-induced resistance to antibiotics in bacteria was approved and funded by the NIH and was transferred from the University of Pennsylvania to Stanford. Cohen arrived in Palo Alto in March 1968 — raring to go.

Cognizant of the collective research prowess of the faculty in the Department of Biochemistry and of the anticipated benefits of working side by side with them — and perhaps even teaching in their courses, Cohen was eager to acquire a joint faculty appointment in biochemistry. But as mentioned earlier, Kornberg was firmly opposed to joint appointments in his department and Cohen's request was viewed no differently than any other such overture over the years — an outcome that left him with something of a bitter taste.

"The biochemistry department faculty was a very distinguished one," he told interviewer Sally Hughes. "But frankly I thought that some of them had an elitist attitude. I was accepted as a Department of Medicine colleague and I believe that they respected my science. But I always felt that they had different standards for evaluating research in a clinical department. I hoped that a joint appointment in biochemistry might be a way of formalizing the scientific relationships that existed *de facto*. But I soon realized that this was not likely to happen."[11]

Predictably, renovation of the research laboratory assigned to Cohen in the Department of Medicine was incomplete when he arrived at Stanford! However, he was rescued by the generosity of biochemistry faculty member David Hogness, who provided Cohen temporary workspace in his own laboratory. This brief sojourn not only avoided undue delay in mounting his research program, but additionally afforded Cohen the opportunity of becoming acquainted with individuals in the biochemistry department. "Most of the faculty were considerably older than me (or at least it seemed that way)," he related. "I viewed the students and postdocs as more my contemporaries."[12]

At that time, many investigators around the world studying plasmids and bacteriophages were intent on transfecting bacterial cells with naked DNA — then an exercise in frustration because bacteria are protected from the environment by both a thin membrane and a more robust outer chitin layer that resists uptake of naked DNA. As mentioned in an earlier chapter, the many investigators struggling with DNA transformation included Peter Lobban, who was then attempting to transfect *E. coli* with phage λ together with naked λ DNA, a procedure often referred to as the "helper assay" and known to improve the efficiency of DNA transfection. But this limitation was resolved in early 1970 by Mort Mandel and Akiko Higa at the University of Hawaii, who made a significant technical breakthrough by demonstrating that incubation of *E. coli* cells in the presence of calcium chloride followed by heating rendered them highly competent to the uptake of naked DNA.

Lobban, who was then in search of a reliable and efficient method of transfection of *E. coli* with native DNA to enrich his own studies, was of course elated to learn of these results. But exacting as he was in the laboratory, after some trial and error Lobban concluded that the transformation conditions described by Mandel and Higa were not optimal for his needs. "After hearing from Janet Mertz about her use of the technique, Cohen, who also had a use for the calcium chloride procedure, often visited my lab while I was doing these experiments," Lobban related. "So we freely exchanged information back and forth and eventually Stan adopted the improved conditions that I was able to work out."[13] Mertz, who then also had a need to learn the Mandel–Higa procedure (see later chapters) was also a frequent visitor to Lobban's laboratory bench, a nuance we'll return to presently.

Cohen unstintingly confirms the open and communicative environment in the department of biochemistry at that time. "Students and postdocs mentored by different faculty in that department frequently occupied benches in the same lab and there was extensive exchange of information among both trainees and the faculty," he related. "That was one of the special features of the biochemistry department."[14]

✱✱✱✱✱✱✱✱✱✱✱✱✱✱✱✱✱✱

As recounted more fully in a later chapter, Paul Berg was awarded a share of the 1980 Nobel Prize in Chemistry for his contributions to the development of recombinant DNA technology. Stanley Cohen and his collaborator Herbert Boyer at the University of California at San Francisco were not so honored, though there is widespread consensus that based on scientific merit alone such largesse would not have been surprising. This disappointment may have aggravated (at least in Cohen's mind) subliminal tension between Cohen and Berg around that time, a gestalt that set the stage for a heightening of these stresses.

Cohen, and of course students and postdoctoral fellows in the biochemistry department in general, were aware that Lobban had begun experimenting with poly(dA) and poly(dT) tailing by terminal transferase in order to achieve end-to-end joining of phage P22 DNA molecules. Though not directly privy to conversations between Lobban and David Jackson about the terminal transferase procedure, Cohen was certainly aware of the explicit acknowledgements to Lobban regarding experiments reported in the Jackson, Symons and Berg paper that appeared in October 1972 issue of the *PNAS*.

Mertz also visited Lobban's laboratory to learn about his improvements of the Mandel–Higa technique. Her recollections of that time were related to the author:

> I learned the Higa and Mandel technique from Peter Lobban in December 1970. I have a page in my notebook with that date that is

a photocopy of a page from Peter's notebook. Peter had improved the method using linear λ phage DNA so that he could use it also with P22 DNA. That winter/spring I showed that it worked as well for λ*dv* circular plasmid DNAs that lacked free ends. *Paul prompted me to inform Stan that the protocol worked for plasmids and that he should try it with his R factor plasmids* [author's italics]. After I told him that the technique worked for plasmids and literally handed him a copy of my protocols for doing so, he too went to Peter to learn how to do the method.[15]

Cohen was therefore presumably aware, if not fully informed, about Mertz's experiments, including her intention to learn the Higa and Mandel procedure for transfecting *E. coli* with plasmid DNA.

Toward the end of 1969, Cohen had established that multiple molecular classes of antibiotic-resistance plasmids (called R-factors) could exist concurrently in *E. coli* as circular DNA molecules. These findings were published in high-profile journals — the last issue of *Nature* in 1969 and some months later the *Journal of Molecular Biology*. Cohen *et al.* subsequently established that novel R-factors could be generated by linkage of discrete units that independently harbored genes with different biological functions, such as antibiotic resistance, autonomous replication, recombination, and plasmid transfer.[16] However, the functional and physical relationships between these putative genes remained unknown. It was to address these questions that Cohen began transforming *E. coli* with plasmid DNA. In mid-1972, he and his colleagues Annie Chang and Leslie Hsu reported:

Treatment of *E. coli* cells with calcium chloride ········ renders them capable of taking up molecules of purified R-factor DNA. Moreover ············ the introduced R-factor DNA can persist in such cells as an independently replicating plasmid, and can express both the fertility and antibiotic resistance functions of the parent R factor.[17]

It wasn't immediately apparent to many people that transformation by plasmid DNA was different from previous genetic transformation, because plasmids were autonomous replicons that could be stably inherited without being integrated into the chromosome. No homology with chromosomal DNA is required. There wasn't much interest in R-factors in general then, and I suppose that the title of our paper:

Nonchromosomal antibiotic resistance in bacteria: genetic transformation of Escherichia coli by R-factor DNA, wasn't especially compelling. Most people didn't grasp the significance of being able to take a plasmid out of a cell, introduce it into another cell, and then select cells that contained plasmids that were the progeny of a single DNA molecule.[18]

In light of Cohen's newfound ability to dissect functional components of R-factor DNA, the capacity to propagate individual plasmids was an important technical consideration. "I hoped to obtain clonal populations of bacterial cells that contained different plasmid DNA entities," he told interviewer Sally Hughes in 2009.[19] To this end, Cohen and his colleagues fragmented plasmid DNA by mechanical shearing, the utility of which was seriously limited by the stochastic nature of this crude technique. Fortuitously, restriction endonucleases were then appearing on the molecular biology scene and may have prompted Cohen to idly reflect about their potential utility for more reliably generating DNA fragments.

By May 1972, the work demonstrating that R-factor DNA can replicate independently as a plasmid in *E. coli* and can express both the fertility and antibiotic resistance functions of the parent R-factor was published in the *PNAS*. However, at that time the potential for using plasmids as vehicles for gene cloning was not yet obvious.

Notes

1. SC interview by Sally Hughes, Regional Oral History Office of the Bancroft Library at the University of California, Berkeley, 1995.
2. *Ibid.*
3. ECF interview with Stan Cohen, September 2011.
4. *Ibid.*
5. *Ibid.*
6. SC interview by Sally Hughes, Regional Oral History Office of the Bancroft Library at the University of California, Berkeley, 1995.
7. Letter from Berg to Hal Holman dated February 25, 1971. Reproduced with permission from the Department of Special Collections, Stanford University Libraries.
8. *Ibid.*
9. *Ibid.*

10. SC interview by Sally Hughes, Regional Oral History Office of the Bancroft Library at the University of California, Berkeley, 1995.

11. *Ibid.*

12. *Ibid.*

13. ECF interview with Peter Lobban, September, 2011.

14. SC interview by Sally Hughes, Regional Oral History Office of the Bancroft Library at the University of California, Berkeley, 1995.

15. E-mail from Janet Mertz, dated May 1, 2012.

16. Cohen SN, Chang AC, Hus L. (1972) Nonchromosomal antibiotic resistance in bacteria: genetic transformation of *Escherichia coli* by R-factor DNA. *PNAS* **69**: 2110–2114.

17. *Ibid.*

18. SC interview by Sally Hughes, Regional Oral History Office of the Bancroft Library at the University of California, Berkeley, 1995.

19. *Ibid.*

PART IV

CHAPTER FIFTEEN

An Historic Meeting in Hawaii

I'll Have the Chopped Liver Please

In 1972, a joint US–Japan meeting in Hawaii to consider recent advances in plasmid biology was organized by Cohen, Tsutomu Watanabe (a Japanese investigator credited with discovering plasmid-borne antibiotic resistance) and Donald Helinski at the University of California at San Diego. Scheduled for three days in mid-November 1972, the meeting fortuitously coincided with the appearance of the November 1972 issue of the *PNAS* containing the papers by Morrow and Berg and Mertz and Davis concerning the utility of the restriction endonuclease *Eco*RI for linearizing circular DNA — and in so doing generating cohesive ends.

The November 1972 issue of the *PNAS* also featured an article by Joseph (Joe) Hedgpath, Howard Goodman and Herb Boyer,[1] reporting the nucleotide sequence of the *Eco*R1 cleavage site. Events had moved so quickly that details of Bob Yoshimori's discovery of *Eco*RI enzyme had not yet been published, though Hedgpath *et al.* alluded to this finding in the introductory paragraph of the November 1972 paper just mentioned. When Watanabe passed away shortly before the meeting Helinski and Cohen invited Herb Boyer to fill this unexpected hiatus in the program. "I didn't know Boyer personally and I knew relatively little about his work, although I had seen some of his papers," Cohen related. "But I knew that he had published a review on restriction enzymes and several other papers in that area."[2]

Microbiologist Stanley Falkow (from whom Cohen had obtained plasmid DNA when Falkow was at Georgetown University in Washington, DC) was now a faculty member at Stanford. A celebrated plasmid authority in his own right, Falkow was also at the Hawaii meeting. About 35 years later (in 2009), he documented his recollections of events at the conference in an article delightfully titled: *"I'll Have the Chopped Liver Please, or How I Learned To Love the Clone: A Recollection of Some of the Events Surrounding One of the Pivotal Experiments That Opened the Era of DNA Cloning."* Much of this historic memoir begs reproduction here. Notably, Falkow attributed comments to both Cohen and Boyer concerning the Mertz and Davis observations on the ability of *Eco*RI restriction endonuclease to generate cohesive ends at sites at which the enzyme cuts duplex DNA.

In November 1972 I went to Honolulu for the first time. When I arrived in — the early afternoon, I ran into [Herb] Boyer. ·······························
······························· Boyer and I spent several hours on the beach drinking Blue Hawaiians and talking about the focus of the meeting, R-factors (actually, the new term was R-plasmids). After dinner, we sat in the lobby again contemplating yet another Blue Hawaiian, when Cohen appeared. Within minutes, Charlie Brinton and his wife Ginger also joined our threesome. ·······················.

Plasmids were a hot topic in 1972. ·····································
·································. At one level, we understood that R-plasmids are composed of DNA, are relatively small in relationship to the bacterial chromosome, and encode multiple antibiotic resistance factors. However, in another context we knew relatively little about the biology of plasmids ·······················. We lacked the biochemical knowledge to understand how these extrachromosomal elements are transferred from cell to cell, how they replicate in a wide variety of bacterial species, and how they mediate antibiotic resistance. At the experimental level, there was a good deal of frustration. The resolving power of bacterial genetics and what passed for molecular biology at the time did not provide enough information, nor did there appear to be a major breakthrough in the offing.

Boyer, Cohen, the Brintons and I set out to find something to eat on a Sunday night in Waikiki. We walked down a dark, quiet road and found a commercial area where, on a corner, a bright sign announced a New York-style delicatessen. We settled into a booth and in the spirit of this extraordinary scene, laughed that the Hawaiian waiter knew the intricacies of the menu but not the pronunciation of its elements. After the first surge of calories [which presumably included chopped liver], a serious discussion began. Cohen began to talk in detail about experiments that he and Lucy Chang had done ·················· [in which they sheared a large transmissible plasmid in] a Waring blender ···················· [and transformed] pieces into *E. coli*, with subsequent genetic selection for one or more of the six antibiotic resistances encoded on the plasmid. This had been possible because the pair had established a reproducible DNA transformation system for gram-negative bacteria [referring to the Higa and Mandel technique]. Remarkably, their transformation system resulted in the isolation of a small, self-replicating plasmid encoding tetracycline resistance. ·· ··· .

Boyer's laboratory had recently discovered a restriction enzyme, *Eco*RI. [This enzyme] had been discovered in 1971 ············· by Bob Yashimori as an R-plasmid-encoded restriction enzyme. Boyer had solicited a plasmid [called RSF1010] ················ from my laboratory as one of the means of investigating the specificity of the enzymatic cleavage properties of the *Eco*RI enzyme. In the deli Boyer described to all a finding that he had revealed to me privately that afternoon: *Eco*RI cleaved RSF1010 only once along the entire chromosome. ·····························

In what seemed an unconnected thought, Boyer recounted Janet Mertz's observation about "sticky" DNA ends and some of the work that Peter Lobban was doing with P22 phage in Dale Kaiser's lab to [perform] enzymatic joining of DNA molecules. The conversation to that point had been animated and often consisted of four people talking simultaneously. However, the juxtaposition of the facts about *Eco*RI, its behavior on small plasmids, that common "sticky ends" were formed by *Eco*RI cleavage, and that plasmid DNA could now be easily genetically transformed, caused a silence to fall over the table. It was much too obvious to slip by unnoticed.

Cohen said in a slow, clear voice, "That means ⋯⋯⋯⋯" Boyer didn't let him finish. "That's right, it should be possible ⋯⋯⋯⋯ "Sometimes in science, as in the rest of life, it is not necessary to finish a sentence or thought. The experiment was straightforward enough. Mix *Eco*RI-cut plasmid DNA molecules and rejoin them and there should be a proportion of recombinant plasmid molecules formed that could be isolated by Cohen and Chang's transformation method. However, the larger implications of the work were not lost on us that evening. The excitement was palpable, and the idea of isolating DNA fragments randomly cleaved with *Eco*RI was quickly obvious. The idea of joining distinct DNA species had been at the cutting edge of molecular biology and was, in fact, the focus of Berg's group, as well as those of Kaiser and Lobban, but here was a direct way to do the experiment. ⋯⋯⋯⋯⋯⋯⋯⋯⋯⋯⋯⋯⋯⋯⋯

In retrospect, it is extraordinary that we didn't discuss this set of experiments again for the next two days. The meeting began the next morning, and our time was taken fully with the realities of the moment, rather than with experiments of the future.⋯⋯⋯⋯⋯⋯⋯⋯⋯⋯⋯⋯ ⋯⋯⋯⋯⋯⋯⋯⋯⋯⋯⋯⋯⋯⋯⋯.

Boyer and I saw each other that evening. He had met briefly with Cohen and they had discussed the logistics of the experiments they planned to perform together upon their return to the Bay Area. Boyer and I agreed to meet the next morning and travel to the airport together. We then spent the next hour or so sitting in a quiet spot discussing the splicing experiment and its implications. He asked what role I wanted to play in the planned work ⋯⋯⋯⋯⋯. It seemed to me that all I had to offer was some purified plasmid DNA and this was not a major contribution that required co-authorship. ⋯⋯⋯⋯⋯⋯⋯⋯⋯⋯ On the way home from Hawaii, *I mulled over what the consequences would be if the experiments that Cohen and Boyer planned to do were successful. I knew that if these experiments worked, the science I practiced would never be the same* [author's italics].[3]

Cohen's recollections of the events that unfolded in Hawaii (recounted to Sally Hughes in 2009, about 35 years after the meeting) are related presently. But he steadfastly maintains that when he arrived on the island he was unaware of the Mertz and Davis findings published in the November 1972 issue of the *PNAS*.

My recollection is that the November 1972 issue of the *PNAS* was not actually out in print until early December, 1972.* But in any case, I did not see the Sgaramella, Mertz and Davis, Morrow and Berg, and Hedgpath *et al.* papers until after returning from the Hawaii meeting and a subsequent short vacation on the Kona coast *and did not know of Janet's and Ron's findings with SV40 prior to going to Hawaii* [author's italics].[4]

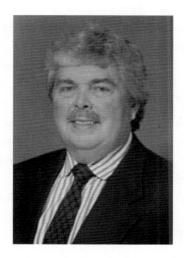

Herb Boyer. (http://web.mit.edu/~invent/a-winners/a-boyercohen.html)

Cohen's contention that he was unaware of Morrow's and Mertz's studies on *Eco*RI enzyme *before* he arrived in Hawaii is very much at the heart of differing opinions on the chronology of events advanced by Mertz and independently by Cohen. In a nutshell, Mertz remains perplexed by Cohen's contention that he "did not know of Janet's and Ron's findings with SV40 prior to going to Hawaii."

"Paul prompted me to inform Stan [Cohen] that the protocol worked for plasmids and that he should try it with his R-factor plasmids. I told him that the technique worked for plasmids and literally handed him a

---

* The physical availability of new monthly scientific journals a few weeks after the end of the month with which they are identified is not at all uncommon. Indeed, it is the rule rather than the exception.

copy of my protocols for doing so. Mertz stated.[5] Then too, the Stanford Departments of Biochemistry and Genetics were in close physical proximity on the third floor of the medical school building, separated only by a hallway. Consequently, information generally diffused readily throughout the floor. Having been invited to present her research at a highly visible Cold Spring Harbor Laboratory meeting on viruses in the fall of 1972, Mertz adhered to the biochemistry department's rule that graduate students and postdoctoral fellows deliver an announced rehearsal talk to an open audience in anticipation of comments and criticisms that might arise during their formal presentation.

In a word, Mertz finds it perplexing that knowing of the ongoing work in the Berg and Kaiser laboratories, Cohen would have been unaware of her and Davis' demonstration that *Eco*RI enzyme leaves cohesive ends at sites where the enzyme cuts DNA — and their contention that two different DNAs bearing the same *Eco*RI-generated ends should be amenable to covalent joining, to yield recombinant DNA — thereby almost certainly facilitating gene cloning.[6]

In the spring of 2012, Mertz sent a communication to the author in which she spelled out her recollections of this historical conundrum in greater detail. In light of the fact that the content of this lengthy and historically relevant missive is in the public domain, it is presented here, if for no other reason than providing documentation for future historians. Mertz's communication states:

> I am absolutely sure Paul told Stan in 1971 that I had shown one could use the Higa and Mandel calcium procedure for reestablishing plasmid DNAs (i.e., λ*dvgal*120) in *E. coli*. Paul then asked me to give Stan my protocol for doing so, so Stan could use it with his R-factor plasmids. I proceeded to do as Paul had asked me to do. That is why Stan went to Peter Lobban to learn the method. In some of Stan's early reviews, he admits knowing I had put λ*dvgal* into *E. coli* with the technique before he used it for his R-factor work, but claims λ*dvgal* was phage DNA containing the *E. coli* gal operon and Higa and Mandel had already shown that the protocol worked for phage λ. The problem with Stan's argument is that λ*dvgal* is a circular plasmid like R-factor plasmids, not a linear DNA with ends as is phage λ DNA. Thus, Paul and I were

informing Stan that we had a protocol that would work with circular as well as linear DNAs. Stan fails to give me proper credit for making this discovery before him. My notebook pages show clearly that I made this discovery in the winter-spring of 1971. ··

I can't prove Stan did not know about the Mertz and Davis results before the 1972 Hawaii meeting on R-factors, since I don't remember who was in the audience when I gave my biochemistry department talk on my findings in August 1972, before going off to CSH [to the meeting to which she had been invited] and in the fall of 1972 at our departmental Asilomar gathering. I just find it extremely hard to believe [Stan] didn't know given [that] lots of folks in the biochemistry department knew: Herb Boyer and Howard Goodman knew at UCSF, Vittorio Sgaramella working with Lederberg in Stan's department knew, the hundreds of folks who attended the August 1972 Cold Spring Harbor meeting on Tumor Viruses knew, etc. AND, according to Stan Falkow, Herb Boyer stated very clearly to Cohen at the Hawaii meeting that the just-published Mertz & Davis *PNAS* article described a protocol for easily joining together 2 plasmid DNAs (i.e., SV40 and *dvgal*) that should be generally applicable for joining ANY two DNAs that contain *Eco*RI sites in them. Given [that] Stan's lab was located just down the hallway from mine and he frequently attended biochem department seminars, it is very hard for me to believe he hadn't heard about my results on how to make recombinants prior to the Hawaii meeting. Even if he hadn't, Boyer clearly told him about the published article at the meeting. Thus, at the time Stan claims he first came up with the idea for joining together 2 R-factor plasmids, Boyer informed him that the already published Mertz & Davis article described exactly how it could be done.

Stan had probably also heard earlier from Paul about our self-imposed moratorium against cloning SV40-*dvgal* recombinants in *E. coli* that had come out of the discussions with Bob Pollack and others in 1971 along with the plans for the upcoming Asilomar I meeting to discuss biohazards. Thus, the idea of cloning recombinant DNAs in bacteria using a plasmid cloning vector also was not a new concept in November of 1972. Stan's new contribution was the idea to use pSC101 instead of *dvgal* as the cloning vector since it encoded a drug selectable marker, making it easier to identify the bacteria that had acquired the plasmid DNAs. ALL of the other techniques and ideas that made Stan's first cloning experiment possible had already been well developed by folks in the biochemistry department with the help of Boyer's *Eco*RI enzyme.[7]

As the original source of *Eco*RI enzyme, Herb Boyer was unambiguously aware of the Morrow/Berg and Mertz/Davis findings in the Berg laboratory. As mentioned earlier, when Berg instructed Mertz and Davis to inform Boyer about their exciting observations, Boyer visited Stanford post-haste to hear about the developments first-hand and to examine the relevant data. During that visit Boyer offered to sequence the *Eco*R1 cleavage site, a body of work that ultimately lead to the previously mentioned Hedgpath *et al.* paper communicated to the *PNAS* by Berg.

Close to 30 years later (in 2001), Boyer also addressed questions about who knew what and when, during an interview by Sally Hughes.

> So we went off to Hawaii. We had done sequence work on the site cleaved by the [*Eco*RI] enzyme. *We knew [about] the work of Ron Davis and Janet Mertz, who worked in Paul's lab at Stanford. We had given some of our EcoRI enzyme to Paul Berg. He gave some to these people* [meaning Mertz and Davis]. *They had shown physically that when the RI enzyme cleaved DNA the ends would reanneal, that they were what we call sticky ends. And this was the same time that we determined the sequence of the cleavage site* [author's italics].
>
> The juxtaposition of the facts about *Eco*RI, its behavior on small plasmids, that common "sticky ends" were formed by *Eco*RI cleavage, and that plasmid DNA could now be easily genetically transformed, caused a silence to fall over the table. It was much too obvious to slip by unnoticed.[8]

What "was much too obvious" was of course the abrupt comprehension that joining linear DNA bearing single-stranded cohesive ends generated by *Eco*R1 enzyme to a different linear DNA bearing the same cohesive ends would be an obvious experimental approach to explore. But there can be no denying that the "obvious" had already been stated in the article in press in the *PNAS* by Mertz and Davis communicated to the journal by Berg.

In 2009, Cohen related his own recollections of the fateful evening stroll with Boyer, Falkow and others in Hawaii, to interviewer Sally Hughes:

Stanley Falkow.
(http://news.stanford.edu/news/2007/june6/med-mcp-060607.html)

Herb Boyer and I discussed the work ongoing in our laboratories. Falkow, Charles Brinton and his wife Ginger and I ended up at a delicatessen (we decided to eat there because the sign said *Shalom* instead of *Aloha*) and continued to talk over sandwiches and beer. Herb initially was not very interested in looking at plasmid genes and offered to provide *Eco*R1 enzyme as a gift for the experiments I wanted to do. He said he had given *Eco*RI to various other people at Stanford [notably John Morrow in Berg's laboratory] and he'd be willing to give some to me. I said, 'Well, that doesn't seem quite fair. Your lab has spent a lot of time isolating the enzyme and we should really do this as a collaboration.' And that's the way we decided to proceed."[9]

Much of this brouhaha derives from the contention by Mertz and others in the Berg laboratory that she and Ron Davis have (to date) been denied appropriate recognition for their priority in predicting the emergence of gene cloning by using a suitable DNA vector cut with *Eco*R1 enzyme. So the question sometimes arises, at least in the mind of Janet Mertz, whether or not Cohen was aware of Mertz's and Davis' observations and the predictions derived from them *before* he attended the plasmid meeting in Hawaii in mid-November or, as Cohen claims, not until he arrived at the meeting. Regardless, it is relevant to note that when in her 2009 interview with Cohen Sally Hughes posed the direct question: "So who first made the discovery that *Eco*RI generates complementary DNA ends that can be joined by [DNA] ligase, Cohen's response was: "*I think that most people credit Mertz and Davis for that finding.*"[10]

Other statements by Cohen to Sally Hughes that bear on this issue warrant mention.

> [T]he original experiments in collaboration with Herb were not specifically designed to clone DNA [author's italics]. They were designed to take a big plasmid with multiple antibiotic resistance genes, treat it with *Eco*RI and allow the assortment of joined fragments. We began with a plasmid of 100kb that had 6 antibiotic resistance genes, chopped it up with *Eco*R1 and deployed the transfection procedure I had developed with Peter Lobban's help to look for recombined fragments by selection of genetic markers we knew about. We found such entities and they contained just three fragments — a region required for replication, a fragment carrying an antibiotic resistance gene, and a random fragment that had come along for the ride.[11]

Since the DNA fragment required for replication was devoid of genetic markers for plasmid selection, Cohen searched for plasmids carrying antibiotic resistance genes that were *not* cut by *Eco*R1 within the antibiotic resistance and replication-determining fragments. "We found one and called it pSC101," he stated. "[*We then*] *realized that we could use this plasmid as a vector for cloning pieces of DNA.* [author's italics]. We initially referred to it as a cloning 'vehicle' because the term 'vector' sounded too mathematical to me!"[12]

> At [the Hawaii meeting] we certainly realized the potential utility for using this method to clone DNA from other sources. But we were really quite cautious about what we said in the [May 1973 *PNAS*] paper, viz, ["*the general procedure described here is potentially useful for insertion of specific sequences from prokaryotic or eukaryotic chromosomes or extrachromosomal DNA into independently replicating bacterial plasmids*"], since

the generality of what we had done was not yet determined. The restriction-modification systems of *E. coli* did not destroy the new *E. coli* plasmid constructs we had made, but we had no idea what would happen if we tried taking DNA from another species and putting it into *E. coli*.[13]

The information Cohen gleaned at the Hawaii meeting concerning the generation of cohesive ends at *Eco*R1 cleavage sites, coupled with the observation that such cleavage sites enabled the joining of two different DNAs, initiated collaborative experiments with Boyer for generating recombinant plasmids. Cohen reasoned that when cut by *Eco*R1 enzyme each of the resulting DNA fragments would likely contain only a few genes, a significant improvement over the mechanical shearing methods he had used earlier for taking plasmids apart. Furthermore, since the ends of the multiple plasmid DNA fragments would be cohesive, the pair anticipated that individual plasmid DNA fragments in the mixture might be joined in different combinations. If, for example, *Eco*R1 cleavage left the replication function of the plasmid intact, the replication region might hook up to different antibiotic resistance genes in the mix, thereby generating DNA circles containing different fragment combinations. "We could then try to genetically transform calcium chloride-treated *E. coli* cells with the ligated DNA and perhaps we could isolate cells containing plasmids with different combinations of antibiotic resistance genes," Cohen related.[14]

Soon after the Hawaii meeting, Cohen and Boyer began testing these predictions. Annie Chang (Cohen's long-time technician) lived in San Francisco and transported DNA back and forth between the Boyer laboratory in San Francisco and the Cohen laboratory at Stanford. Cohen's lab sent Boyer plasmid DNA to treat with *Eco*RI enzyme and then returned it to Cohen. Cohen's lab in turn performed DNA transformations and genetic selection and again forwarded the DNA to Boyer, who analyzed it by gel electrophoresis. The results convinced Boyer that he and Cohen had hit pay dirt. "… *when I looked at those gels I knew we'd be able to isolate any piece of DNA that was cut with EcoRI, regardless of where it came from* [author's italics]," he told interviewer Jane Gitschier in September 2009.[15]

As soon as the generation and isolation of what ultimately became the widely used cloning vector pSC101 were completed, Cohen and Chang addressed the fate of "foreign" DNA in this plasmid. To their satisfaction they observed that DNA from a different bacterial species (Staphylococcus) also resided perfectly happily in pSC101. The Discussion section of this landmark paper entitled "Genome Construction Between Bacterial Species *in vitro*: Replication and Expression of Staphylococcus Plasmid Genes in *Escherichia coli*," published in the *PNAS* in April 1974 (17 months after the Hawaii meeting), states:

> It may be practical to introduce into *E. coli* genes specifying metabolic or synthetic functions (e.g., photosynthesis, antibiotic production) indigenous to other biological classes. *In addition, these results support the earlier view that antibiotic-resistance plasmid replicons such as pSC101 may be of great potential usefulness for the introduction of DNA derived from eukaryotic organisms into E. coli, thus enabling the application of bacterial genetic and biochemical techniques to the study of eukaryotic genes* [author's italics].[16]

Years later, Cohen told interviewer Sally Hughes:

> The discovery that a Staphylococcal gene could be transplanted to *E. coli* and be propagated there surprised a lot of people. The earlier experiments involved genes isolated from *E. coli*, but this was DNA from an unrelated organism. Even though we had suggested the potential general utility of these methods in our earlier paper, the work with the Staphylococcal plasmids provided evidence that we could actually use these methods *to clone foreign DNA in E. coli.* [author's italics].[17]

Regardless of nuances about who predicted or pronounced what and when, the significance of the studies by Cohen and his colleagues during the late 1960s and early to mid-1970s, sometimes in collaboration with Boyer, are unassailable. Berg categorically agrees with the assessment that while his laboratory was the first to make contributions to the emergence of recombinant DNA technology, Cohen *et al.* were the first to experimentally extend this breakthrough to gene cloning.

As already mentioned, Berg received the Nobel prize in Chemistry in part for his contributions to the development of recombinant DNA technology. Some in the scientific community not unreasonably contend that Cohen and Boyer were (are) equally deserving of such recognition. However, the fact that the award is limited to three individuals in any given year, coupled with the Nobel Foundation's 1980 decision to honor both Wally Gilbert and Fred Sanger for their contributions to DNA sequencing technology, meant that only a single individual could be so honored for contributions to recombinant DNA technology, a not uncommon conundrum when Nobel prizes are handed out. As for Janet Mertz and Ron Davis, the indignant sentiments expressed by Mertz concerning "appropriate recognition" are certainly understandable. But as just stated, it is commonplace for deserving individuals, especially those in the very early stages of their careers as scientists, to experience this sort of disappointment.

A final word by Peter Lobban on the topic of gene cloning bears reflection: "I find it remarkable that *anyone* in the biological sciences would not immediately see the desirability of being able to isolate pieces of DNA and attach them to something that would ensure their replication. Perhaps the reason many people back then seemed not to comprehend this had something to do with the compartmentalization of the biological sciences into biochemistry, biophysics, genetics, and so on."[18]

Notes

1. SC interview by Sally Hughes, Regional Oral History Office of the Bancroft Library at the University of California, Berkeley, 1995.
2. Hedgpeth J, Goodman HM, Boyer HW. (1972) DNA Nucleotide Sequence Restricted by the R1 Endonuclease. *PNAS* **69**: 3448–3452.
3. Falkow S. (2001) I'll Have the Chopped Liver Please, or How I Learned To Love the Clone: A Recollection of Some of the Events Surrounding One of the Pivotal Experiments That Opened the Era of DNA Cloning. *ASM News* **67**: 555–559.
4. E-mail from Stanley Cohen dated April 25, 2102.
5. ECF interview with Janet Mertz, February 2012.
6. ECF interview with Janet Mertz, February, 2012.

7. E-mail correspondence with Janet Mertz, February 14, 2012.

8. HB interview by Sally Hughes, Regional Oral History Office of the Bancroft Library at the University of California, Berkeley, 2001.

9. SC interview by Sally Hughes, Regional Oral History Office of the Bancroft Library at the University of California, Berkeley, 2009.

10. *Ibid.*

11. *Ibid.*

12. *Ibid.*

13. *Ibid.*

14. *Ibid.*

15. HB interview by Jane Gitschier, September 2009. Wonderful Life: An Interview with Herb Boyer. *PLoS Genet*, September 2009, e1000653.

16. Chang ACY, Cohen SN. (1974) Genome Construction Between Bacterial Species *In Vitro*: Replication and Expression of Staphylococcus Plasmid Genes in *Escherichia coli*. *PNAS* **71**: 1030–1034.

17. SC interview by Sally Hughes, Regional Oral History Office of the Bancroft Library at the University of California, Berkeley, 2001.

18. E-mail from Peter Lobban, September 3, 2012.

CHAPTER SIXTEEN

The Recombinant DNA Controversy

The gathering storm

Almost as soon as the Jackson, Symons and Berg account of generating SV40/λ*dvgal* recombinant DNA appeared in the *PNAS* of October 1972, red flags began waving. A *Nature* biology correspondent was the first to address a commentary on the published work:

> What would be the consequences if the reagent Berg and his colleagues have made somehow infected and lysogenized *E. coli* in someone's gut as the result of an accident? This possibility, remote though it may seem to be, can hardly be ignored, and it will be most interesting to learn what criteria the group adopts when it decides whether or not the scientific information that might be obtained by continuing the experiment justifies the risk.[1]

Stanford University public information director Spyros Andreopoulis had a standing arrangement with the medical school's library that brought all journal references to experiments involving Stanford faculty to his prompt attention. On receiving a copy of *Nature's* commentary on the Berg experiment, Andreopoulis visited Berg and asked whether the time might not be opportune for a statement to the press that would bring the hybrid experiment into the open and possibly avoid a serious misunderstanding later. Berg's response was a blast of indignation — directed not at Andreopoulis but at *Nature*. He said the editors of the journal had known that the part of the experiment criticized by the cell biology correspondent had been abandoned and had denied him due credit for recognizing a problem and acting to eliminate it. Andreopoulis argued for a clarifying letter to the editor of *Nature*. Berg declined to entertain the suggestion.[2]

Nor was this the earliest alarm bell. Unease about the experiments surfaced as early as the summer of 1971 in the midst of the fledgling efforts by David Jackson. As mentioned earlier, during that summer Janet Mertz returned to the East Coast to attend a workshop on cell culture at the Cold Spring Harbor Laboratory on Long Island, New York. Robert (Bob) Pollack, a Cold Spring Harbor Laboratory scientist was in charge of the course. Mertz's application to win a place in this highly subscribed event was enthusiastically endorsed by Berg, who wrote to Pollack: "You would not regret having her as a student since she [will] not only liven up the group, but she [will also] provide a spark that could easily rub off on the other students."[3]

Robert (Bob) Pollack.
(www.earth.columbia.edu/articles/view/2738)

When Mertz arrived at Cold Spring Harbor she discovered that she was one of but two graduate students present, the other attendees being either postdoctoral fellows or faculty members. "When we got her application it was clear that we couldn't turn her down because she had A's in everything, had graduated from MIT in three years, and had a bang-up letter from Paul Berg," Pollack related. "So we accepted her, but with many misgivings, because the course required a certain minimal maturity that permitted you to think of yourself as a student, even though you had the self-confidence to know that you also had a degree and were a practicing scientist. People who did best in these courses were those with the most self-confidence; who could enjoy being told what to do for a change. Someone like Janet, who was terribly sensitive, self-conscious and defensive about whether she would be accepted into the course or not went bananas when she was constantly told what to do."[4]

"The course was a three-week affair with a ton of material to be covered," Mertz related. "One learned laboratory techniques each morning. Every afternoon we had a journal club where we discussed papers from the literature and we had seminars in the evening. It was no vacation — every day was very busy."[5]

One afternoon, Pollack led an open discussion on biological safety in the laboratory. During the ensuing dialogue Mertz raised the issue of experiments that might pose health risks. Never shy to volunteer opinions, she suggested that most safety issues in the laboratory could surely be satisfactorily addressed by using appropriate equipment (such as laminar flow hoods) and laboratory technique (such as not pipetting by mouth).

But the events that transpired in Cambridge, Massachusetts on the heels of Beckwith's and Shapiro's press conference were still fresh in Mertz's memory and her thoughts turned again to her dilemma about experiments that might eventually lead to genetic engineering in humans. She raised this issue during the open discussion led by Pollack by explaining the technology that the Berg laboratory had in mind for making recombinant DNA. "I told them that all the technology other than that needed for actually generating recombinant DNA molecules was already available in Paul's laboratory and that we were almost at the point where we should be able to amplify mammalian DNA in *E. coli* and retrieve it for use in a variety of experiments."[6] Transmittal of this information to the assembled conferees must certainly have opened the already leaking floodgates about the pending arrival of tools for gene cloning in the Berg laboratory.

Sometime in the next 24 hours, presumably soon after he had fully digested the implications of Mertz's comments, Pollack approached the young Stanford graduate student — his face wracked with concern. "Do you really mean to put a human tumor virus into *E. coli* — a gut bacterium?" he asked incredulously. "Don't you know that SV40 is a tumor virus in some mammals?"

"My initial reaction was that this was utter nonsense," Mertz stated. "We would be working with a laboratory strain of *E. coli* that would have a hard time surviving outside that special environment. And if one was truly concerned about spreading cancer genes one could easily generate additional mutations in the bacterium to render it absolutely harmless outside the laboratory. This wasn't a novel concept. Bacterial geneticists generated mutations in bacteria all the time for all sorts of experiments."[7] Mertz was in fact suggesting a containment procedure that was later independently surfaced by others much more her senior, at a seminal meeting at Asilomar, California in February 1975.

Pollack's brief recollections of this situation are related in an interview for the MIT Oral History Program in 1976. "Janet had a terribly aggressive air about her, which was not satisfied by compliments, unlike most people's aggressiveness," he related. "She was bragging about how they were going to make 'Russian dolls': SV40 in λ, λ in [*E*] *coli*. I told her: 'Well it's *coli* in people!' She didn't know that as far as I could see. It never occurred to her."[8]

Pollack was acutely tuned to ethical and moral issues in science. In addition to his position as professor of biology at Columbia University, he joined the Earth Institute there in 2009 as director of the Center for the Study of Science and Religion.[9] Having taught microbiology at New York University prior to his sojourn at Columbia, Pollack was also well versed in the biology of SV40 virus. He knew that the virus could provoke tumor formation in mice and transform human cells cultured in the laboratory to a state that resembled cancer in certain respects. But frank neoplasia had never been documented in humans, even when the Salk polio vaccine was found to be tainted with SV40 virus. Additionally, Pollack had long been uneasy about the *laissez-faire* manner in which large quantities of SV40 virus were being propagated by investigators at the Cold Spring Harbor Laboratory, where Jim Watson had assembled an animal virus research group.

As the course neared its conclusion, Pollack invited Berg to deliver the closing lecture. Berg declined. For one thing he had no desire to travel across the country to give a single lecture in someone else's course, nor did he especially appreciate the late notice. So the lecture was delivered by Pollack himself and bore the evocative title: "Our Research in Introducing Foreign DNA — Is It Anybody Else's Business Beside the Scientists Who are Doing the Work?"[10]

In the final days of the course, Pollack and Joseph (Joe) Sambrook, a resident tumor virologist at the Cold Spring Harbor Laboratory, composed a lengthy memorandum on the topic at hand that was distributed to the course participants. Most of the memo was not especially noteworthy. But according to historian John Lear, the last segment posed the question: "*Are there any good experiments using human cells and viruses that should not be done?*"

Even if you chose to, you could not experiment on human beings. You all accept, therefore, that a boundary exists past which you must not let your curiosity carry you. Work with somatic human cells in culture has proceeded without any apparent hesitation, so such work defines the inner or "doable" side of this boundary. Now a class of experiments with human germ cells (e.g., *in vitro* fertilization, cloning, introduction of "absent" genes, or gene therapy) is becoming possible.

Will you place those experiments on the same side as experiments on skin-biopsy cells, or will you group them with experiments on people? You have mobilization time to consider this question *before* you begin the work.

A second related class of experiments involves the reverse process: putting human genes or the nucleic acid of human viruses into cells of other species, or into prokaryotic cells. The ethical problem here is minor, but the dangers (e.g., of creating a human tumor virus that can grow inside a bacteria like *E. coli* which normally sits in the human gut) are immense.

If you are going to work with either of these two classes of experiments, we suggest you surrender a portion of the scientist's right to follow his nose without regard to consequences. You ought to ask yourself if the experimental results are worth the calculable dangers. You ought to ask if the experimental techniques are over the boundary and amount to experimentation with people.

> Finally, you ought to ask yourself if the experiment *needs* to be done, rather than if it *ought* to be done, or if it *can* be done. If it is dangerous, or wrong, or both, and if it doesn't need to be done, just don't do it. This is not censorship. You must accept a physician's responsibility if, by free choice, you work within these classes of experiments.[11]

Mertz has no recollection of this memo. "We received dozens of hand-outs of one sort or another, so it may have been lost in the shuffle," she explained.[12] As the end of the course approached, the then-21-year-old Stanford student was in a quandary. On the one hand, she was convinced that the work she was engaging in Berg's laboratory (generating a plasmid containing SV40 DNA to introduce into *E. coli*) would yield an excellent PhD thesis. But on the other hand, she realized that the ultimate outcome of what she astutely identified as a "looming controversy" was beyond her control, and in the final analysis decisions might be made that would preclude her continuing some if not all her planned experiments. "Concerns about the moral and ethical implications of genetic engineering might still have been lingering in the back of my mind," Mertz related. "But mainly my perspective was that I was a lowly graduate student and this sort of politics was completely out of my league. I also believed that even if my present long-term goal of expressing SV40 genes in *E. coli* was scotched, I had lots of other good ideas for a PhD thesis!"[13]

In her March 1977 interview for the MIT Oral Program, Mertz recounted her concern and anxiety in those early days.

> Coming from a radical background, I figured, "Well, even if there's only a 1 in 10^{30} chance that there's actually something dangerous that could result, I just don't want to be responsible for that type of danger." I started thinking in terms of the atomic bomb and similar things. I didn't want to be the person who went ahead and created a monster that killed a million people. Therefore, pretty much by the end of the week I had decided that I wasn't going to have anything further to do with this project, or, for that matter, with *anything* concerned with recombinant DNA.[14]

After the Cold Spring Harbor course, Mertz continued to mull on this issue, and in her agitated state she decided to call Berg on the telephone. When she related the disturbing events of the previous weeks, Berg was generally dismissive. "I didn't think that Paul had thought about it much," Mertz said, "So I tried to sound a warning note to him by telling him that there were people out there who were upset and worried. 'You'd better start thinking about this,' I said."[15] Berg's recollection of that phone conversation is essentially that the more counter arguments he raised, the more counter-counter arguments Mertz raised in return!

When she returned to Stanford, Mertz was distraught. "I pretty much told Paul that I wasn't going to work on trying to introduce SV40 into *E. coli* any longer," she stated. "So at that point I started working on something else. I began to examine the infectivity of various forms of SV40 DNA, including linear molecules and relaxed DNA circles. By August 1971 John Morrow had shown that the *Eco*RI restriction enzyme cut SV40 once at a unique site and I was soon embroiled in showing that in so doing it generated cohesive ends in the cut DNA."[16]

At that time, concerns about more invidious risks of recombinant DNA had not been publicly expressed. Concern was largely (if not exclusively) focused on the hypothetical cancer risk associated with introducing SV40 DNA into *E. coli*. Apprehension about more worrying — possibly even terrifying outcomes of genetic engineering would emerge later as the controversy heightened.

On June 28, 1971, shortly before Mertz's return to Stanford, Pollack too contacted Berg by telephone. He did so hesitantly. Pollack was well aware of Berg's stellar scientific reputation. He had in fact grown up in Brooklyn a mere few blocks from the Berg residence and, like his own parents, Pollock's parents fondly referred to Berg as Paulie. But having never met the man as an adult, he was concerned that the Stanford biochemist might have an indignant reaction to anyone poking their nose into his scientific business.

To Pollack's relief Berg was not obviously affronted. He had already heard about Pollack's concerns from Mertz and had also begun privately mulling about the issue. But his conversation with Pollack mimicked that with Mertz in the sense that each party raised arguments, counter-arguments and counter-counter arguments. "I remember Pollack specifically saying that he felt strongly that these experiments shouldn't be done — and I was really quite annoyed by that," he stated indignantly.[17] But later reflection led him to the wise decision to canvas scientific colleagues whose opinions he respected about the ominous safety question that had surfaced. He met with Josh Lederberg at Stanford — who viewed the risks that Pollack had raised as highly improbable. But like others before him, Lederberg confronted Berg with the obvious impossibility of absolutely ruling out these and possibly other as yet unimagined risks.

Berg also recalled a visit to the NIH around that time, during which he met with a number of virologists. "The Berg experiment scares the pants off a lot of people, including me," Dr. Wallace Rowe, a NIH virologist, is quoted as stating in the journal *SCIENCE*.[18] Another specialist in virology, George Todaro of the National Cancer Institute (NCI) at the NIH, affirmed that the experiment was "one of those I think just shouldn't be done."[19] Berg also met with his long-time friend and scientific colleague Maxine Singer, who directed a research laboratory at the NIH. Author John Lear commented: "Dr. Maxine Singer was very critical of the experiment in spite of her warm personal friendship with Berg."[20]

Maxine Singer and her husband Daniel (Dan), a lawyer with strong opinions about moral and ethical situations in science not infrequently invited Berg to dine at their home when he was in Bethesda. He received such an invitation during one of his visits to the NIH to solicit views about his experiments. At that dinner, Berg was introduced to Leon Kass, a staff fellow at the NIH who served as executive secretary to the *Committee on the Life Sciences and Social Policy* from 1970–72 and in later years chaired the President's Council on Bioethics.[21]

Kass and Dan Singer were friends and both were associates of the *Hastings Institute of Society, Ethics and the Life Sciences*, a bioethics think-tank located close to New York City. Kass had earned an MD degree from the University of Chicago and a PhD in biochemistry and molecular biology from Harvard. Early in his career he began directing his research to the ethical and social implications of advances in biochemical science — and the possible consequences of genetic manipulation. Indeed, fears of nightmare scenarios (such as generating unimaginable life forms in the laboratory — and even interfering with biological evolution) may already have begun to surface. If not, they did soon thereafter.

Berg's recollection of the dinner conversation at the Singer residence is that he maintained a defensive position about the ethical validity of the experiments going on in his laboratory. "I was pretty much convinced that many of the fears that people were expressing were red herrings and that they had such low probabilities, that while one could state them as possibilities, the likelihood of anything like that happening was extremely remote."[22] A short time after this dinner, Kass sent Berg a lengthy letter with a copy to Maxine and Daniel Singer, in which he recapitulated their dinner conversation and raised a formidable list of issues that he felt merited further discussion. Nonetheless, he began his lengthy epistle by congratulating Berg on his openness and sensitivity to ethical questions. "It was a great pleasure to meet and talk with you," Kass wrote, "Your sensitivity to the possible ethical and social implications of your own work gave me much encouragement."[23]

But the more discussions he entertained with others, the more Berg realized that while he could cogently argue that the probability of unpleasant consequences was indeed remote, he could never claim that it was zero. "I realized that I'd been wrong before in predicting the outcome of my experiments," he stated. "But if I was wrong about the risk factors associated with using recombinant DNA the consequences were not something I wanted to live with for the rest of my life."[24] In fact, Berg recalled having a conversation with postdoctoral fellow David Jackson that led to considerations of forgoing trying to introduce SV40 DNA into bacteria and concentrating exclusively on experiments designed to introduce recombinant DNA into mammalian cells, thereby taking the contentious *E. coli* experiments off the table.[25] Regardless, as stated earlier, when Mertz joined the laboratory in early 1971, the notion of trying to express mammalian genes in *E. coli* had resurfaced.

In his 1975 interview for the MIT Oral Program on Recombinant DNA, Berg related another debate he had with himself — and with Jackson, essentially assessing whether there was any point in grappling with the many technical issues that might surround transduction of mammalian DNA into *E. coli* if in fact mammalian genes could *not* be expressed in bacterial hosts. This consideration was particularly relevant given the fact that having successfully introduced SV40 genes into *E. coli,* the bacteria would be carrying potential oncogenes that may conceivably find their way into human cells. "In other words, we might be doing experiments that would have no scientific payoff — but would still carry risks," he stated.[26]

The red flags proliferated! Berg recalls a late night debate (with much beer in attendance) with a number of young and vocal participants that transpired on the roof of an old castle overlooking the Straits of Sicily. This "all-nighter" unfolded at a NATO Summer Course held from July 27–29, 1971 in Erice, Sicily. That year, the course, an annual event organized by Francis Crick and others, included Berg as an invited lecturer on a topic unrelated to recombinant DNA technology. But after complying with a request to provide an additional less formal presentation on his recombinant DNA work, Berg received an outspoken response from the younger participants. "We stayed up late, drinking beer and discussing the possible hazards of and prospects for genetic engineering," he stated.[27] But John Lear's lyrical statement: "With the ghost of Adolf Hitler still goose-stepping among them, they [the young scientists] saw themselves poised on the brink of a frightening new era of human engineering and behavior control,"[28] is perhaps overly dramatic!

Though by no means personally convinced about the risk of cancer associated with his experiments, Berg ultimately came to the decision that it would be unwise to continue this work in the midst of so much adverse opinion — and he so informed members of his laboratory. "[At] that early stage I decided to put our plans for testing the biological properties of the recombinant molecules in bacterial cells on hold until we could better assess [their] health risks," he later wrote.[29] "So, we set aside testing whether SV40 λ*dvgal* DNA could grow in *E. coli* (the experiment that Berg had earlier suggested to Mertz). We had also intended to introduce the plasmid into mammalian cells, but David Jackson had already left the laboratory — so that experiment too was put on hold"[30] In a nutshell, Berg voluntarily (though not necessarily willingly) instituted a moratorium on any recombinant DNA experiments in his laboratory.

When Berg informed Pollack of his decision to discontinue work that involved introducing recombinant DNA into *E. coli* until further notice, the Cold Spring Harbor scientist was naturally pleased. Still, having been sensitized to possible risks associated with working with SV40 (and possibly other potentially harmful viruses) in the laboratory, Berg decided that it might be opportune to assemble a group of relevant scientists to discuss health risks generally associated with laboratory-based research that utilized viruses — omitting considerations of recombinant DNA for the time being. He suggested that Pollack organize such a meeting. Pollack was in fact genuinely admiring of Berg's responsiveness. "I had put him on the spot, because he's an honest guy and he didn't have an answer to my question," he stated. "But typical of the man, he didn't just give me a brush-off; he used judo on the whole argument — just flipped the whole thing over and instead of being the guy who is the *problem*, he became the guy who is ostensibly the *solution*."[31]

Aided by several knowledgeable colleagues, Pollack invited about 100 participants to discuss biohazards potentially linked to viral research. The meeting, supported by the NIH and the National Science Foundation, convened at the Asilomar Conference Center, a charming and tranquil facility tucked away in the splendid Pacific Grove area south of San Francisco and a venue long used by the Stanford Department of Biochemistry for its annual scientific retreat. The assembled conferees met from January 22–24, 1973. This meeting is often referred to as "Asilomar I," to distinguish it from the previously mentioned larger, more focused and certainly more contentious meeting (Asilomar II) convened about two years later at the same venue.

Four session topics were identified for the Asilomar I meeting: "*Laboratory Infections Introduced by Experimental Animals or Animal Cell Cultures*"; "*Endogenous Agents in Cell Cultures: Detection, Elimination and Risk*"; "*The Tumor Viruses Themselves*" and "*Hazards Associated With Modern Research Methodology*." The last-mentioned session was restricted to presentations on arboviruses and hepatitis virus, existing methods for the large-scale production of tumor viruses, immunosuppression and cancer, and Kuru and the viral dementias. "It was a very useful experience in that it brought in one place information known not only about tumor viruses, but also the use of animal cells in culture and the potential risks they brought forth,"[32] Berg related.

The one hundred conferees included David Baltimore, Jim Watson and Norton Zinder, all destined to play prominent roles in subsequent developments. But little of real consequence emerged from Asilomar I. A small book entitled *Biohazards in Biological Research* (edited by Alfred Hellman, Michael N. Oxman and Bob Pollack) was published by Cold Spring Harbor Laboratory Press. Few in the scientific community were aware of, let alone read, the book. "It seems safe to say that most laymen would not find *Biohazards in Biological Research* worth the $5 charged by the Cold Spring Harbor Laboratory for a copy of the book," John Lear wrote.[33]

"People came away from the meeting recognizing, as I did, that it was inappropriate to be cavalier about the hazards of working with viruses in the laboratory, because in point of fact we had no basis — no experimental way — of verifying whether or not risks were in fact associated with this type of work," Berg stated. "But a major suggestion that came out of that meeting was that the NIH should undertake a prospective epidemiological experiment by gathering data from individuals working with tumor viruses, i.e. collect a database so that if in the future there was a significant increase of some particular pathology amongst that group one would at least have real data to go back to. This turned out to be more complicated that we naively thought. The NIH informed me that they would have great difficulty in devising a questionnaire or some other intelligence gathering device that would not upset a lot of people — or that recommendations that might emerge from such studies would be adhered to."[34]

The judgment not to address possible dangers of recombinant DNA technology at the Asilomar I meeting was in no way prompted by a calculated decision to be evasive. The primary risk factor then under consideration was the (remote) possibility of a cancer epidemic associated with exposure to SV40 virus and other cancer-causing viruses. This despite the fact already mentioned that known contamination of the Salk oral polio vaccine by SV40 virus was not associated with any documented evidence of increased cancer risk. But neither Berg nor Pollack were then contemplating the nightmare scenarios that would ensue when the emergence of recombinant DNA technology became more apparent — and ultimately reached public ears and eyes.

In concluding this chapter, it bears repeating that while the published (and ongoing) studies on recombinant DNA emanating from the

Stanford laboratories were not broadly communicated, the work in Berg's laboratory was never a calculated secret. Without question many at Stanford knew what was going on in his laboratory and some were (presumably) not reticent to communicate this news to others. Nor was Berg especially close-mouthed when presenting research seminars and talking to scientists around the world, such as the conversation that transpired at the NATO workshop in Erici mentioned earlier. Additionally, as he had promised before departing the Salk Institute at the conclusion of his sabbatical year, he made frequent reports back to colleagues there about his progress in recombinant DNA research. Consequently, the word spread through the scientific grapevine and individual scientists began contacting Berg to learn more — and predictably, to seek advice in spreading their own wings in the recombinant DNA winds. As early as beginning May 1970, Berg received a letter from Richard Roblin, then at the Salk Institute, requesting information about Berg's recombinant DNA work:

> Several people who have visited Salk recently from Stanford have mentioned a research project going on there involving an attempt to link *lac* operon DNA and SV40 DNA as a means of constructing a molecule which might integrate foreign DNA into human cells. I thought this sounded interesting and promising and that I would attempt to find out a few more details by writing to you.[35]

Then too, word had spread from the Cold Spring Harbor Laboratory following Bob Pollack's course and must surely have attracted interest. Indeed, another missive to Berg received in early September 1970 came from Heiner Westphal, a member of the scientific staff at Cold Spring Harbor Laboratory, who wrote:

> David Zipser and I [are thinking] about trying to integrate SV40 DNA into *E. coli* and [have discussed] several ways to achieve this. However, Joe [Sambrook] just told me that you have a student working on a similar project, involving polyoma DNA. Therefore, before we decide whether to begin such a project we would like to make sure that we are not repeating work done in your laboratory.[36]

In late 1971, Berg received an invitation from British virologist Michael (Mike) Fried at the Imperial Cancer Research Fund in London to "come and tell [us] about the state of your 'engineering' projects." Berg responded: "It seems that what we have to do is to get some more work done and stop talking about what we plan to do!"[37]

Notes

1. *Nature* **240**: 73, 1972
2. Lear J. (1978) *Recombinant DNA: The Untold Story*. Crown Publishers Inc., p. 51.
3. Letter from Berg to Robert Pollack, March 15, 1971. Reproduced with permission from the Department of Special Collections, Stanford University Libraries.
4. RP interview by Mary Terrall, MIT Oral History Program, March 1976.
5. JM interview by Mary Terrall, MIT Oral History Program, March 1977.
6. *Ibid.*
7. *Ibid.*
8. RP interview by Mary Terrall, MIT Oral History Program, March 1976.
9. http://www.earth.columbia.edu/articles/view/2738
10. Lear J. (1978) *Recombinant DNA: The Untold Story*. Crown Publishers Inc., p. 26.
11. *Ibid.* p. 27.
12. ECF interview with Janet Mertz, February 2012.
13. *Ibid.*
14. JM interview by Mary Terrall, MIT Oral History Program, March 1977.
15. ECF interview with Janet Mertz, February 2012.
16. *Ibid.*
17. ECF interview with Paul Berg, April 2012.
18. Lear J. (1978) *Recombinant DNA: The Untold Story*. Crown Publishers Inc, p. 37.
19. *Ibid.*
20. *Ibid.*
21. http://www.aei.org/scholar/leon-r-kass/
22. Interview with Paul Berg, September 2011.
23. Kass letter to Berg, October 23, 1971. Reproduced with permission of the Department of Special Collections, Stanford University Libraries.

24. PB interview by Rae Goodell, MIT Oral History Program, May 1975.
25. *Ibid.*
26. *Ibid.*
27. *Ibid.*
28. Lear J. (1978) *Recombinant DNA: The Untold Story.* Crown Publishers Inc, p. 37.
29. Berg P. (2008) Moments of Discovery. *Annu Rev Biochem* **77**: 15–44.
30. PB interview by Sally Hughes, Regional Oral History Office of the Bancroft Library at the University of California, Berkeley, 2000.
31. RP interview by Rae Goodell, MIT Oral History Program, March 1976.
32. PB interview by Rae Goodell, MIT Oral History Program, May 1975.
33. Lear J. (1978) *Recombinant DNA: The Untold Story.* Crown Publishers Inc, p. 55.
34. PB interview by Rae Goodell, MIT Oral History Program, May 1975.
35. Letter from Richard Roblin to Paul Berg dated May 7, 1970. Reproduced with permission of the Department of Special Collections, Stanford University Libraries.
36. Letter from Heiner Westphal dated September 4, 1970. Reproduced with permission from the Department of Special Collections, Stanford University Libraries.
37. Letter to Mike Fried dated November 29, 1971. Reproduced with permission of the Department of Special Collections, Stanford University Libraries.

CHAPTER SEVENTEEN

A Momentous Gordon Research Conference

Well, now we can put together
any DNA we want to

The Gordon Research Conferences, well known to generations of American and foreign scientists, were initiated by Neil Gordon, an American chemist who in 1931 invited a select group of colleagues to spend a week on an island in the Chesapeake Bay to contemplate "scientific frontiers of the future."[1] Gordon conscientiously continued these annual meetings, ultimately establishing a summer conference format now managed by a huge administrative infrastructure that orchestrates dozens of such annual meetings on topics ranging from nucleic acids to environmental nanotechnology. Some Gordon conference topics have been in existence for decades.

For many years, the conferences convened exclusively at boarding schools in the Northeastern United States vacated during the summer months, an arrangement of financial benefit to both the schools and the conference organization. As their popularity grew, the Gordon Conference organization spread further afield, even crossing international borders. But regardless of whether one is housed at a boarding school in New Hampshire, or in "digs" at Queen's College, Oxford University, the conferences are notorious for their Spartan accommodations, a "tradition" made slightly more palatable by the time-honored lobster banquet on the last night!

At the very outset, Gordon decreed that the subject matter presented at these conferences was to be strictly confidential. To this day it is forbidden to quote anything stated at a Gordon Conference in the scientific literature, or to cite such statements as reference sources. Attendance is typically limited to about 200 conferees (including multiple speakers), a good number of whom are graduate students and postdoctoral fellows. By tradition, scientific sessions convene in the mornings and at night. Afternoons are kept free for informal communication — and relaxation. The conferences enjoy a reputation as one of the premier science venues in the world, and an invitation to present one's research at a Gordon Research Conference is a coveted distinction.

Participants at the Gordon Research Conference on Nucleic Acids in June 1973. Conference chair Maxine Singer and vice-chair Dieter Söll are in the middle of the first seated row. (www.lifesciencesfoundation.org/events)

On June 10, 1973 about 150 conferees convened at the New Hampton School in New Hampton, New Hampshire for a conference on

The Biology of Nucleic Acids under the chairmanship of Maxine Singer from the National Institutes of Health and vice-chairmanship of Dieter Söll from Yale University. Stanley Cohen was not at the 1973 Gordon Conference on Nucleic Acids. However, Herb Boyer was invited to present a talk in a session on *Bacterial Restriction Enzymes and Analysis of DNA*. Berg also did not attend the meeting, John Morrow being the sole representative from his laboratory.

When conferring with Cohen about what might be stated in public and what might best be left unsaid in Boyer's presentation, the pair agreed that it was appropriate to mention the experiments that yielded plasmid pSC101, findings then in press in the *PNAS* under the heading "Construction of Biologically Functional Bacterial Plasmids *in vitro*." But work demonstrating that plasmids could support the replication of foreign (Staphylococcal) DNA in *E. coli* was not yet even in manuscript form, let alone published, so the two scientists agreed that Boyer would not discuss that work during his presentation.

With the exception of Boyer, John Morrow and perhaps a handful of other conferees who were well informed about the experiments ongoing in the Berg, Kaiser and Cohen laboratories at Stanford, few of those in attendance knew much if anything about the events of the past nine months — and no one then realized that gene cloning was literally around the corner. After all, the field of nucleic acids research embraces much more than joining bits of DNA!

Accounts vary about precisely how much of the proverbial cat Boyer let out of the bag during his presentation and the subsequent reactions of the conferees. In a 1975 interview for the MIT Oral History on Recombinant DNA, Maxine Singer stated:

> On the Thursday morning (the penultimate day of the meeting) we had a session on *Bacterial Restriction Enzymes and the Analysis of DNA*. Dan Nathans chaired this session and Herb Boyer was the first speaker. He talked about the *Eco*RI enzyme and its utility for generating recombinant DNA. He also discussed work that he and Cohen had done with drug-resistance plasmids and the transfer of drug resistance to smaller plasmids using *Eco*RI to generate cohesive ends. In that session,

someone, I believe it was Bill Sugden* probably in reference to Herb's talk, prefaced his own presentation with an idle remark to the effect: *"Well then, now anyone can put together any DNA we want to."* It was a remark that I know many people noted since he was obviously reflecting the general excitement ignited by Boyer's presentation. I remember sharing in this enthusiasm. But it wasn't until people came up to me that night, specifically Ed Ziff and Paul Sedat (who I viewed as thoughtful and smart people) and later Sherman Weismann and Dan Nathans, and pointed out the implications of what Herb and Stan were doing, that I realized there was another side to this story and that there was a lot to be concerned about. And of course as soon as they raised this issue I remembered the discussion we had when Paul and Leon Kass were at my home for dinner.[2]

Having worked as postdoctoral fellows with Fred Sanger at the MRC Laboratory of Molecular Biology at about the same time, Ed Ziff and Paul Sedat knew one another well. "We talked about this issue a lot," Ziff stated. "This was a time when essentially nothing was known about the health risks of working with recombinant DNA. It was also a time when health risks were being examined for handling dangerous biological reagents in general, such as oncogenic viruses. Paul had very strong opinions about the correct way of going about things such as this, and about public safety in general," Ziff stated. "He was concerned that people might dive into recombinant DNA experiments without understanding what risks were involved. So after that session we approached Maxine."[3] Singer subsequently related:

Sherman [Weismann] and Dan [Nathans] had real reservations about bringing this matter to the attention of the entire meeting, based chiefly on the fact that we didn't really have much time to do this on the last morning ···················· and we would probably not be able to give the matter the sort of attention it deserved. They also expressed concern

---

* Bill Sugden was then a recently graduated young molecular biologist from Columbia University who was scheduled to present a talk immediately following Boyer's presentation. Sugden, who went on to a distinguished career as a molecular virologist at the University of Wisconsin, has no recollection of this episode.

about discussing something that might lead to restrictions on research. Nonetheless, I announced to the meeting that we would have a brief general discussion the following morning. After that evening session I went back to my room and drafted a written statement to distribute the following morning.[4]

As for Boyer's recollections of the open discussion immediately following his presentation, in 1975, a time reasonably proximate to the Gordon Conference, he told MIT interviewer Rae Goodell:

> .... as I remember it, there were standard questions about various aspects of the experiments. But the business of the potential dangers of recombinant DNA molecules didn't arise during the discussion immediately following the presentation. After the session, however, people were talking amongst themselves for the rest of the meeting.[5]

Regardless of the disparity between several accounts to various interviewers years after the conference, some of the data that Boyer shared with his audience that June morning in 1973 indisputably ignited a sense of excitement in some — and alarm in others. There is also no question that off the cuff remarks by one or more conferees prompted the issue of risks potentially associated with recombinant DNA technology to surface during that scientific session — and to bubble over more volubly after the session concluded. That day certainly marks the beginning of widespread concerns about the possible "evils" of genetic engineering.

The letter drafted by Maxine Singer and signed by both her and Dieter Söll was presented to the entire meeting on the last morning of the conference:

> First I will describe briefly the question that has been raised by some participants in the conference, as I see it. We all share the excitement and enthusiasm of yesterday morning's speaker who pointed out that the scientific developments reported then would permit interesting experiments involving the linking together of a variety of DNA molecules. The cause of the excitement and enthusiasm is twofold. First, there is our fascination with an evolving understanding of these amazing molecules and their biological action, and second, there is the idea that

such manipulations may lead to useful tools for alleviation of human health problems. Nevertheless, we are all aware that such experiments raise moral and ethical issues because of the potential hazards such molecules may engender. In fact, potential hazards exist in some of the viruses many of us are already studying. Other problems will arise with hybrid molecules we are contemplating. Furthermore, these hazards present problems to ourselves during our work and are potentially hazardous to the public. Because we are doing these experiments, and because we recognize the potential difficulties, we have a responsibility to concern ourselves with the safety of our coworkers and laboratory personnel as well as with the safety of the public. We are asked this morning to consider this responsibility.

I fully understand that I have not discussed this topic exhaustively, and that there are even arguments to be made about the factual content of my statement. However, having the problem raised so late in the meeting requires that we deal expeditiously with this question. As Chairman, I will not permit substantive discussion of the problem, but only proposals concerning possible action or inaction on the issue. In fifteen minutes the discussion will be closed and we will vote by a show of hands on any proposals that have been made and seconded. We will then proceed with the full scientific program planned for this morning.[6]

Only 95 of the 147 Gordon Research Conference participants were in attendance on the final day of the conference. A spirited discussion ensued and ultimately a formal motion was approved to address a letter of concern to the *US National Academy of Sciences* and its affiliate the *Institute of Medicine*, and to publicize the letter in the journal *SCIENCE*. When Singer called for a vote on this motion, 70 of the 95 participants at the meeting voted "yea."

Not wanting to exclude those who had left the conference before this special meeting, Singer composed an independent epistle to them in which she reproduced the statement distributed at the special session and offered the choice of approving or not the letter intended for distribution. She stipulated that failure to submit a response would be interpreted as supportive of such a letter. Ultimately, 122 of the 142 original meeting participants supported the measure. Suddenly and dramatically, recombinant DNA technology was poised to take center stage in the annals of biomedical research. In her 1975 interview for the MIT Oral History Project Singer related:

> In the final analysis I was dissatisfied with the situation at the end of the conference. I felt that whatever came out of the entire hullabaloo was flawed by the fact that we had not had the opportunity to have any kind of substantive discussion. On the other hand I believed that if we had done nothing there would have been no serious follow-up. The entire issue would have petered out. Imagine what would have happened if someone like Josh Lederberg who [even at that early stage] was frankly opposed to the possibility of government regulations that might shut down recombinant DNA research, at least temporarily, had been in my position. The discussion, brief as it was, would never have transpired.[7]

As soon as he returned to Stanford John Morrow presented the Singer–Söll letter to Berg — who in turn immediately wrote to Singer:

> John [Morrow] told me of the discussion concerning "potential hazards" and I read your remarks to the Conference on that score. As you must know, I have some interest in the same subject. ·····································.
> I think a study committee of the type suggested would be worthwhile and could clear the air. The only feature of your letter to which I take exception is the "feel" being transmitted that this kind of work may very well be hazardous, rather than what I prefer to think, [which] is [that] we have no way of knowing whether it is or not. The phrase in the first paragraph, "certain of these hybrid molecules *are* potentially hazardous" seems a bit strong, although admittedly the word "potentially" is intended to convey doubt; but the doubt seems borderline. May I offer a substitute for the first two sentences of the second paragraph for your consideration? (We are concerned at the possibility that such hybrid molecules could be hazardous to laboratory workers and the public. Though no indications of such a hazard presently exist, a majority of those attending the Conference voted to communicate their concern to you and to the President of the National [Institute] of Medicine ---------------------------.[8]

Philip Abelson, the editor of *SCIENCE*, received Singer and Söll's letter with an explanatory preamble: "Those in attendance at the 1973 *Gordon*

Research Conference on Nucleic Acids voted to send the following letter to Philip Handler, president of the National Academy of Sciences and to John R. Hogness, president of the National Institute of Medicine. A majority also desired to publicize the letter more widely." The historic letter reads:

> We are writing to you, on behalf of a number of scientists, to communicate a matter of deep concern. Several of the scientific reports presented at this year's Gordon Research Conference on Nucleic Acids (June 11–15, 1973, New Hampton, New Hampshire) indicated that we presently have the technical ability to join together, covalently, DNA molecules from diverse sources. Scientific developments over the past two years make it both reasonable and convenient to generate overlapping sequence homologies at the termini of different DNA molecules. The sequence homologies can then be used to combine the molecules by Watson–Crick hydrogen bonding. Application of existing methods permits subsequent covalent linkage of such molecules. This technique could be used, for example, to combine DNA from animal viruses with bacterial DNA, or DNAs of different viral origin might be so joined. In this way new kinds of hybrid plasmids or viruses, with biological activity of unpredictable nature, may eventually be created. These experiments offer exciting and interesting potential both for advancing knowledge of fundamental biological processes and for alleviation of human health problems.
>
> Certain such hybrid molecules may prove hazardous to laboratory workers and to the public. Although no hazard has yet been established, prudence suggests that the potential hazard be seriously considered.
>
> A majority of those attending the Conference voted to communicate their concern in this matter to you and to the President of the Institute of Medicine (to whom this letter is also being sent). The conferees suggested that the Academies establish a study committee to consider this problem and to recommend specific actions or guidelines, should that seem appropriate. Related problems such as the risks involved in current large-scale preparation of animal viruses might also be considered.
>
> Maxine Singer
> Dieter Söll[9]

Ed Ziff in turn communicated a quick letter to the popular British science magazine *New Scientist*, a periodical not especially widely read in the molecular biological community. But in contrast to *SCIENCE* magazine, which offered no editorial comment on the Singer–Söll letter, the *New Scientist* provocatively stated:

> The article by Dr. Edward Ziff ⋯⋯⋯⋯⋯ is important and signifi-
> cant in two respects. First, there is the seriousness of Dr. Ziff's warn-
> ing to his fellow biologists that newly discovered techniques of DNA
> hybridization might pose threats not only to laboratory workers but also
> to the public.[10]

The *New Scientist* editorial never got around to identifying the second "important and significant issue." Regardless, this was likely the first occasion on which "threats to the public" were raised by any media source. The piece also surfaced the alarming specter of "fabricating ever nastier BW [biological warfare] agents" by genetic engineering. But to its credit it also offered hope about anticipated benefits of "the arrival of the long-heralded science of 'genetic engineering.'"[11]

Notes

1. Lear J. (1978) *Recombinant DNA: The Untold Story*. Crown Publishers Inc, p. 66.
2. ECF interview with Maxine Singer, February 2012.
3. ECF Phone interview with Ed Ziff, ⋯⋯⋯.
4. ECF interview with Maxine Singer, February 2012.
5. MS interview by Rae Goodell, MIT Oral History Project, May 1975.
6. *Ibid.*
7. *Ibid.*
8. Letter from Paul Berg to Maxine Singer, June 1973. Reproduced with permission of the Department of Special Collections, Stanford University Libraries.
9. *SCIENCE* **181**: 1114, 1973.
10. *New Scientist* **60**: 237, 1973.
11. *Ibid.*

CHAPTER EIGHTEEN

Making Recombinant Molecules with Frog DNA

I told Stan that John Morrow had Xenopus DNA

Having concluded his research on restriction mapping of the SV40 genome to the satisfaction of his PhD thesis committee, John Morrow was close to completing writing his dissertation. He had secured a postdoctoral position with developmental biologist Donald (Don) Brown at the Carnegie Institution for Science (formerly the Carnegie Institution of Washington) and was planning to leave California as soon as his thesis was formally approved.

Donald (Don) Brown.
(http://carnegiescience.edu/news/carnegie's_donald_brown)

Among other areas of research, Don Brown was interested in the fibroin gene, expressed in the posterior silk glands of the silkworm *Bombyx mori*. "John Morrow came to visit with me from California while still a graduate student and we talked about various projects," Brown related in his 1976 interview for the MIT Oral History Project. "I knew of the restriction mapping work he was doing in Paul's laboratory so I suggested that he treat some of our frog DNA with *Eco*RI enzyme and take a look at it when he was on the electron microscope again. We had DNA carrying 5S ribosomal genes. ·· I gave him some and sent him back to Stanford."[1]

In the interim, Morrow acquired postdoctoral financial support from both the Helen Hay Whitney and Cystic Fibrosis Foundations, and once ensconced in Brown's laboratory he hoped to purify silk gland DNA, join it to the λ*dvgal* plasmid vector generated by Doug Berg, Janet Mertz and David Jackson and introduce the recombinant DNA into *Xenopus* oocytes. Once back in Berg's laboratory, Morrow established that *Eco*RI restriction endonuclease cut ribosomal frog DNA at a single site, yielding two large fragments, an observation with profound subsequent experimental outcomes — and more than a little fallout.

Soon after returning from the November 1972 meeting in Hawaii, Boyer and Cohen, with help from Cohen's technician Annie Chang, joined forces to consummate the collaborative experiments described earlier, in the hope of isolating *E. coli* cells containing plasmids with different combinations of antibiotic resistance genes (see Chapter 15). As mentioned earlier, Boyer subsequently told interviewer Jane Gitschier that the results firmly convinced him that he and Cohen had hit pay-dirt — they now knew how to clone genes — essentially any genes.[2]

****

This is an apt time to revisit the historic 1973 Gordon Research Conference in New Hampton, New Hampshire. John Morrow arrived at the meeting venue on Sunday, June 10, 1973. His recollections are that he went directly to the registration desk to obtain a key to the room assigned to him and to collect his conference package. Whilst milling around the crowded conference registration area he spotted and joined

Herb Boyer and Bob Helling, also braving the horde of scientists register-
ing for the meeting.

At some point during the conference, Morrow and Boyer entertained
a telling conversation. Morrow believes that this dialogue transpired on
the Sunday preceding the beginning of the meeting, when he spotted
Boyer at the registration desk. "Boyer told me that he and Stan Cohen
were excited about a new plasmid that they had recently generated [refer-
ring to plasmid pSC101]. I recall my distinct interest in that bit of news
in light of my own plans for working in Brown's laboratory."[3]

> Boyer also mentioned that he and Stan were contemplating inserting
> some sort of eukaryotic DNA into their plasmid vector to see whether
> it could replicate in *E. coli*. At that time people were leery about putting
> known cancer-related genes into bacteria. But frog 5S ribosomal DNA
> was not associated with cancer — at least not to my knowledge, so I
> wasn't concerned about safety issues. And I knew that the DNA was cut
> by *Eco*R1 and yielded good size products — something I had already
> informed Don [Brown] about.[4]

So Morrow blithely informed Boyer that he happened to have frog DNA
in his freezer at Stanford and that he had already determined that it was
cut by *Eco*R1 enzyme. The pair excitedly agreed "to do the obvious
experiment when we returned to the Bay Area," namely, ask whether a
plasmid containing *vertebrate* (frog) DNA could replicate and proliferate
in *E. coli* — in more specific terms, ask whether one could clone vertebrate DNA in *E. coli*.

Whether by a trick of memory about events that transpired 38 years
ago or not, Boyer recalls talking to Morrow sometime *after* Boyer's presentation, which John had attentively listened to of course. "I told him
that the obvious next step would be to try and clone fragments of a
eukaryotic genome into pSC101 and see how the construct behaved in
E. coli," Boyer related.[5]

> John immediately informed me that he had frog DNA at Stanford and
> that he knew it was cleaved by the *Eco*RI endonuclease. He offered to
> provide some of it — assuming of course that Don Brown would be

agreeable. (Notably, Berg's agreeability was apparently not considered then.) At that time I assured John that if he wanted to get involved with the experiments he was most welcome. He replied, "I don't think I can. I've got to finish my thesis — and so on and so forth." When I got back to San Francisco I called Stan Cohen and Bob Helling and we all went up to Stanford to plan the experiments. I informed Stan that John Morrow had Xenopus DNA and he knew that *Eco*RI cut it. So we contacted John, who once more steadfastly agreed to provide frog DNA. I again extended an invitation to him to join us as a collaborator — and he again declined. But the very next day he called me and said: "Well, maybe I *will* get involved."[6]

Stan Cohen related his own recollection of these events:

> When Herb returned to his lab he phoned me to ask if I thought that *Xenopus* DNA would be a good eukaryotic DNA to try to clone, and if I did, to invite my participation. Don Brown gave permission for us to use his *Xenopus* DNA in these experiments and after the Gordon Conference John, Herb and I got together and mapped out a strategy for cloning this DNA and identifying hybrid molecules.[7]

The group did not possess the means for selecting bacterial cells harboring recombinant plasmids containing *Xenopus* DNA. But they confidently anticipated that if such DNA molecules were indeed generated, they would be able to determine whether or not they contained frog DNA by deploying electrophoretic techniques previously used to identify Staphylococcal DNA in *E. coli*. "I thought that we probably would have to screen a lot of cell clones for recombinant plasmids, but it was worth a try," Cohen stated.[8]

The experiment went flawlessly and the group reported these exciting results in a paper published in the *PNAS* entitled "Replication and Transcription of Eukaryotic DNA in *Escherichia coli*." The paper (communicated by Josh Lederberg) was published in May 1974. Morrow was the first author. As he was still officially a graduate student in Berg's laboratory, no one would have raised eyebrows had Berg been listed as a co-author. He was not! Many years later, Janet Mertz proffered: "Stan might have believed that Berg was appropriately left off the authorship of the Morrow *et al.* article because John had already obtained his PhD — even though he had not yet left Stanford. The fact is that Morrow was still officially a graduate student working in Berg's laboratory when he was doing the experiments for the Morrow and Berg paper."*[9]

The circumstances by which Berg eventually gained knowledge of Morrow's collaboration were not at all pleasing to him. Morrow maintains that he had earlier informed Berg about his acquisition of Xenopus DNA from Don Brown. Berg has no such recollection. Regardless, upon returning from the Gordon Conference, Morrow made no specific mention to Berg of the impending experiments with Cohen and Boyer. He was then not exactly in favor with his PhD mentor for other reasons. Notably, Berg had been nagging Morrow for the past several months about the imperative of completing his PhD thesis — which in Berg's considered opinion was seriously lagging.

Above and beyond this petty annoyance, Berg was incensed that Morrow had not personally informed him about, let alone asked his permission for his collaboration with Cohen and Boyer. Thirty-seven years later, Morrow needed no reminding of this incident. "It wasn't right of me not to have asked Paul's permission," he forthrightly admitted. "But like lots of situations that reach a boiling point, events evolved gradually. Paul and I had agreed that I had done enough experiments for my thesis. All I had to do was write it. So I was able to justify to myself that it was okay to use time that wasn't committed to work in Paul's lab for other things. And in fact I *was* [author's italics] working on my thesis."

"Paul and I have never tried to reconstruct precisely what transpired week by week at that time," Morrow continued. "As I remember, I began the experiments when he was on vacation that fall. (Berg confirms that

---

* When this book was nearing completion, Berg received an archived letter from Mertz concerning the disposition of some of her personal archives. In the letter Mertz referred to "pages of the abstract booklet from the August 1972 Cold Spring Harbor Laboratory Meeting on Tumor Viruses at which Dave Jackson, John Morrow, and I all presented talks on our work that appeared in PNAS that fall. The letter confirms that you were listed as a co-author on our abstract that reports the findings we later published as Mertz & Davis."

that may well have been the case.) But my overall recollection of working with Paul is really golden. It has a warm, yellow hue about it. Paul was a most wonderful mentor." Elsewhere, Morrow was quoted as saying: "His [Berg's] students aren't merely tools to accomplish research. They are associates he's interested in. Others told me that I should not have done what I did without Paul's permission. In retrospect I fully agree — and have told him so."[10]

"He was contrite beyond belief," Berg commented — with a chuckle. "Even now, 38 years later, John expresses obvious contrition about that incident whenever I see him."[11]

Around the time of this drama, the biochemistry department was preparing to hold its annual scientific retreat at the usual Asilomar venue. Essentially all the faculty, postdoctoral fellows and graduate students in the department attended this event, an eagerly anticipated annual occurrence, especially by trainees slated to present their work. Besides, it was not all work; there was plenty of time for play — and Asilomar was by no means an unpleasant place to spend a few days!

At the height of his anger, Berg had informed Morrow that he was not welcome to attend the retreat! But once his fury cooled, he rescinded this directive and telephoned Morrow inviting him to come to Asilomar forthwith and present his exciting studies. Morrow did so — to considerable acclaim from the assembled faculty — including Berg and Arthur Kornberg. Recombinant DNA technology had now definitively progressed to gene cloning — and the excitement was general.

****

Though not undeserved, Berg's passionate response to John Morrow's indiscretion was by no means exceptional. Generally amiable, pleasant and courteous to all, Berg has a highly developed sense of decorum when dealing with others. But when rubbed the wrong way he has little reservation in revealing his ire, regardless of the perceived status of the recipient. Though Berg's occasional intolerance of others is far from a conspicuous feature of his personality, it has surfaced frequently enough over the course of his career that it merits mention.

In the fall of 1973, Berg and Francis Crick corresponded about a rather trivial issue that in due course became heated. A letter from Berg about SV40 minichromosomes had apparently piqued Crick's interest, as was Berg's intention. In mid-September 1973, Crick wrote: "We would really be most grateful if you could send us a few 'typical' pictures [of the minichromosomes in the electron microscope] with the scale accurately marked, so that we could really see for ourselves."[12]

On October 19 of that year, Berg replied to this letter, politely apologizing for "the delay in sending these pictures." This missive prompted a mild (for the ever forthright Francis Crick) "complaint" expressing his "disappointment not to receive your exciting photos of SV40 chromatin via the Kornbergs." Crick concluded his epistle with an undeniably but not overtly provocative paragraph:

> We are still hoping to receive some of your pictures (with dimensions marked on them), but if you and [Jack] Griffith [an American DNA electron microscopist with whom Berg was collaborating] feel you cannot release them at this stage I would be glad if you would let me know as soon as possible, as in that case we shall have no choice but to repeat the work independently.[13]

Many would find little or no offence in Crick's statement. But Berg felt otherwise and on November 5, 1973 (Guy Fawkes Day in the UK), he lit his own firecrackers when writing again to Crick:

> Although not especially noted for it, I can be aroused to heights of anger, particularly when I believe I've been dealt with unfairly. So if my comments seem too blunt, bear in mind that they originate from the impetuosity of the moment and are not likely to reflect a lasting feeling.
>
> What has provoked me is the implicit and not too subtle threat in the last paragraph of your letter of October 19. After all, Francis, I didn't owe you anything — information or pictures; supplying both was mine and Jack's choice — freely given with the hope and expectation that you and Roger [Kornberg] would find it helpful in your own work. I was especially pained to discover that you (I trust that the thought never entered Roger's mind) believed (even if momentarily) that I was being devious and trying to withhold information from our exchange.[14]

Berg also once (at least) crossed swords with the other half of the double helix. The reaction to Watson's position on Asilomar II has already been mentioned and will be revisited shortly (see Chapter 22). But years later, Berg was incensed when, after having been asked by Watson to comment on a chapter on recombinant DNA in a near final draft of a forthcoming book called *DNA — The Secret of Life*, co-authored by Watson and Andrew Berry, he read the published version. Despite Berg's earlier offering of a lengthy list of corrections and misstatements, the book appeared unchanged from the version Berg was sent. None of the misstatements had been corrected or explained.

One of the comments in Watson's and Berry's manuscript to which Berg took particularly strong exception concerned the order of events — and by implication priority for the discovery of recombinant DNA technology. *"While Cohen and Boyer, and by now others, were ironing out the details of how to cut and paste DNA molecules, Berg planned a truly bold experiment."* Berg was angry about the obviously mistaken chronology implicit in Watson's supposition, which suggested that the work by Cohen and Boyer *preceded* that from his own laboratory, a suggestion that he had specifically indicated to Watson was inaccurate when reading the draft sent to him. He was particularly disturbed that Watson's comment was precisely of the sort that could lead to scurrilous conclusions about Berg's relationship with Peter Lobban at the time that they were pursuing similar strategies to launch recombinant DNA technology. Accordingly, Berg sent Watson an irate communication:

> *You certainly know that our experiments were in progress during 1970–1971 long before anybody was even able to "cut and paste DNA molecules."* [author's italics] ·· Moreover, it was the concern about our experiments raised by Bob Pollack that triggered the first Asilomar Conference in 1973 [which Watson attended]. At that time the Hawaii "deli lunch" at which Cohen and Boyer supposedly conceived their plasmid joining experiments was a year away from happening. Consequently, your presentation indicating that we began our experiments after others were already making and cloning DNA is patently false ················[15]

As quick as he sometimes is to taking exception to the utterings and/or written comments by others, Berg is equally swift in issuing apologies and is certainly not one to bear grudges. In this context, it merits mention that he has demonstrated remarkable sensitivity to the plight of total strangers who over the years have written to him in the hope of hearing encouraging news about impending cures from genetic disorders. He has responded to all such letters identified in his Stanford archive, expressing sympathy in hand-written replies. A typical such letter reads:

> Your letter, like so many others I receive, moved me greatly. I sympathize deeply with the problem your son's illness poses for you and your wife and, of course, to your son. I fervently hope that his vision will improve and that in time he'll be able to live a nearly normal life.[16]

Notes

1. DB interview by Rae Goodell, MIT Oral History Project, May 1976.
2. HB interview by Jane Gitschier, September 2009. Wonderful Life: An Interview with Herb Boyer. *PLoS Genet*, September 2009, e1000653.
3. ECF interview with John Morrow, September 2011.
4. *Ibid.*
5. HB interview by Jane Gitschier, September 2009. Wonderful Life: An Interview with Herb Boyer. *PLoS Genet*, September 2009, e1000653.
6. *Ibid.*
7. SC interview by Sally Hughes, Regional Oral History Office of the Bancroft Library at the University of California, Berkeley, 1995.
8. *Ibid.*
9. ECF interview with Janet Mertz, February 2012.
10. ECF interview with John Morrow, September 2011.
11. ECF interview with Paul Berg, September 2011.
12. Letter from Francis Crick to Paul Berg, September 18, 1973. Reproduced with permission from the Department of Special Collections, Stanford University Libraries.
13. Letter from Francis Crick to Paul Berg, ············ 1973. Reproduced with permission from the Department of Special Collections, Stanford University Libraries.

14. Letter from Paul Berg to Francis Crick, November 5, 1973. Reproduced with permission from the Department of Special Collections, Stanford University Libraries.
15. Letter from Paul Berg to James D. Watson, April 12, 2003. Reproduced with permission from the Department of Special Collections, Stanford University Libraries.
16. Reproduced with permission from the Department of Special Collections, Stanford University Libraries.

CHAPTER NINETEEN

The Controversy Heats Up

I would not do that experiment now!

Whold Philip Handler (President of the US National Academy of Sciences) received the Singer–Söll letter at the conclusion of the June 1973 Gordon Conference on Nucleic Acids, he forwarded it to a subcommittee of the academy recently created by him called the Assembly of Life Sciences. At its first meeting the Assembly astutely decided that it required expert scientific opinion about what was going on and how to proceed further, and turned to Maxine Singer, who in turn volunteered

Phillip Handler.
(http://www.nasonline.org/about-nas/history/archives/presidents)

Berg as the best source of such information. Berg was in the midst of discharging his teaching obligations at Stanford when Handler called to ask him to help advise the Assembly of Life Sciences as to how to respond to the issues raised in the Singer–Söll letter. But Berg assured Handler of his firm intention to convene a group of scientists with experience in this area to discuss how best to proceed. "I wasn't concerned about having to bear sole responsibility as much as I believed that comprehensive discussions by the group would likely provide the best advice to Handler," Berg related.

> At that stage, I had no idea of what was to come. I was just completing my term as department chairman and was eagerly anticipating getting more in touch with what was going on in the lab. Indeed, Janet [Mertz] was teaching me electron microscopy of DNA and I was thoroughly enjoying the seclusion of the electron microscope facility in the basement of the Beckman Center. I naively believed that the advice of the group would be passed on to Handler, and once it was in his hands I would be free of additional involvements.[1]

On April 17, 1974, David Baltimore (from MIT), Dan Nathans (from Johns Hopkins University), Jim Watson (from Cold Spring Harbor Laboratory), Sherman Weissman (from Yale University), Norton Zinder (from Rockefeller University), Richard Roblin (from the Salk Institute) and Berg convened at MIT as an informal working group — herein simply referred to as the MIT Committee. Each of the members was cognizant of the events surrounding the June 1973 Gordon Conference and the purpose of the meeting convened on the final morning of that conference. They were also familiar with the paper announcing the construction of the "cloning vehicle" pSC101 by Stan Cohen, Herb Boyer and their colleagues, which appeared in the November 1973 issue of the *PNAS*, and knew that a follow-up paper by Chang and Cohen reporting that plasmid pSC101 containing Staphylococcal DNA could also be propagated in *E. coli* was to appear in the April 1974 issue of the *PNAS* — within days of the MIT meeting.

Berg notified his MIT colleagues of the impending publication of the experiments by Morrow *et al.* demonstrating the propagation of plasmid pSC101 containing eukaryotic (*Xenopus*) DNA in *E. coli*. He also informed them that Stan Cohen was receiving a flood of requests for his plasmid and related biological reagents from his laboratory. Clearly, if the MIT Committee was to make any action decisions — they required urgent attention!

Cohen was not at all pleased with his newfound "popularity." He was in the midst of the most exciting work of his career and was loath to be distracted by issues of protocol and scientific collegiality. "Stan was really quite upset," Berg related. "He didn't know how to respond, particularly since when he asked people why they wanted plasmid pSC101, they sometimes described experiments that others had cautioned him about. I was really shocked too, because in many ways they were exactly the kinds of experiments that we had been forewarned about a few years earlier — experiments that involved cutting up herpes virus DNA to put into *E. coli* — and that sort of thing. Stan was reticent to send out plasmids and strains and I advised him to hold off until I returned from the MIT meeting."[2]

By early 1974, Berg had himself taken delivery of multiple requests for advice about initiating recombinant DNA experiments in their own laboratories. One was from a microbiologist, Patricia Berg (no relation) at the University of Chicago. The gravity with which Paul viewed this letter is evident from his response, dated February 11, 1974:

> Dear Dr. Berg,
> I have a great many reservations about the experiment you propose and a great deal of thought should be given to possible hazardous consequences. .... My own prejudice (and it is only a prejudice at present) is to call a moratorium on such experiments ....
> I can't expect that my prejudice is enforceable, but you did ask my opinion. And that is, I would not do that experiment now![3]

The MIT Committee became increasingly disturbed about the pace at which events seemed to be unfolding and were in full agreement that some means was urgently required to convey the need for caution to scientists most likely to deploy the new cloning technology. Ultimately the group converged on the advisability of a letter signed by all, to be published in one or more science journals that would *suggest* (rather than *dictate*) that people consider voluntarily deferring certain experiments until after their potential risks were better understood. Secondarily, the group focused on the imperative of convening an international conference to thoroughly and thoughtfully examine the recombinant DNA issue — soon!

By pure coincidence, Cohen happened to be visiting MIT to present a research seminar the very day that Berg's committee met. He became aware of the intended letter to *SCIENCE, Nature* and the *PNAS* and upon returning to Stanford he asked Berg to add his name to the list of seven signatories, pointing out that none of them had particular experience with plasmids. Berg acceded to this request (though not necessarily solely because he thought that the letter required the signature of a plasmidologist) — and in the interests of keeping "all relevant parties happy" he added Herb Boyer, Ron Davis and David Hogness as further signatories. The Assembly of Life Sciences of the US National Academy of Sciences approved the report on May 24, 1974 and the letter appeared in the July 26, 1974 issue of *SCIENCE* [4] and subsequently in *Nature* and the *PNAS*.

When the letter was received by National Academy chairman Phil Handler for his endorsement, Handler was distinctly irate because of his growing impression that the academy was essentially serving as "a free post office" for letters on their way to publication in scientific journals; first the Singer–Söll letter and now the Berg letter. He insisted that the "Berg letter" (as it came to be known) be identified as a formal report from the *Committee on Recombinant DNA Molecules, Assembly of Life Sciences, National Research Council, National Academy of Sciences*. The entire content of this historic letter is not reproduced here in the interests of brevity.

To Berg's chagrin, the letter is sometimes referred to as the "moratorium letter," implying that he and his committee unilaterally dictated a halt to all recombinant DNA work without consulting members of the scientific community at large. But in fact, the word "moratorium" does not appear in the published letter. "We never called for a total cessation of recombinant DNA work," Berg explained. The letter stated:

> The undersigned members of a committee, acting on behalf and with the endorsement of the Assembly of Life Sciences of the National Research Council on this matter, propose the following recommendations:
>
> First, and most important, that until the potential hazards of such recombinant DNA molecules have been better evaluated or until adequate methods are developed for preventing their spread, scientists throughout the world should join with the members of this committee in voluntarily deferring the following types of experiments.

The letter went on to describe two categories of experiments (Type I and Type II) that should be deferred, and listed three further recommendations. It ended with the following plea;

> .... our concern for the possible unfortunate consequences of indiscriminate application of these techniques motivates us to urge all scientists working in this area to join us in agreeing not to initiate experiments .... until attempts have been made to evaluate the hazards and some resolution of the outstanding questions has been achieved.

The published letter had an international impact. The UK Medical Research Council (MRC) requested a full halt to recombinant DNA experimentation until a formal report by a special committee was thoroughly examined and discussed. The Japanese, Russians, most of the Europeans and scientists from other countries agreed to observe the requests made in the *SCIENCE* letter. Additionally, Berg was asked to make a personal appearance at the Royal Institution in London (not to be confused with the Royal Society) and another at a meeting at the Karolinska Institute in Stockholm, both venues designed to shore up arguments supporting the recommendations in the *SCIENCE* letter. "I ended up doing considerably more traveling and speaking that I was intending," Berg related, "and the impact on my research program was becoming noticeable and troubling!"[5]

The journal *SCIENCE* rapidly became a favored repository for numerous opinions on the increasingly emotionally charged topic, particularly those contributed by Nicholas Wade (then a science correspondent for the *New York Times*). Wade's published comments and opinions continued well into the 1970s, generating considerable heat from Berg, who was of the considered opinion that as its name implied, *SCIENCE* was primarily a *scientific* journal and should not serve as a convenient forum for *non-scientists* to express opinions. The July 26, 1974 issue of *SCIENCE* included one such communication from Wade. The letter began:

> Because of a remote but possible hazard to society, a group of molecular biologists sponsored by the *National Academy of Sciences* has called for a temporary ban on certain kinds of experiments that involve genetic manipulation of living cells and viruses. This is believed to be the first time, at least in the recent history of biology that scientists have been willing to accept any restrictions on their freedom to research, other than those to do with human experimentation.[6]

Wade reported opinions from a number of scientists, several of whom expressed their frustration that recombinant DNA experiments either already ongoing or in their future plans, were clearly to be delayed — if not banned outright. He pessimistically concluded his lengthy commentary on a cautionary note:

> Quite possibly the embargo will be observed until the [Asilomar] conference in February. Its real test will come when and if the conference decides the hazard is substantial enough for the embargo to be indefinitely extended. It could then become apparent that control of the new technique is not much easier than the containment of nuclear weapons.[7]

In early 1977, *SCIENCE* published yet another commentary from Wade, prompting Berg to communicate directly with the Editor-in-Chief, Philip Abelson:

> I am curious about how the decision is made as to where Nicholas Wade's articles belong. His contributions of the past year or so could

hardly be considered news reports. ···
·····························. I have followed the field of recombinant
DNA research carefully for some years now, but have failed to uncover
the scientific facts that justify the deductions Wade made in his January
29th article entitled "*Dicing With Nature: Three Narrow Escapes.*" To my
knowledge there is not a single scientific fact to warrant Wade's implication that the world was brought to the brink of disaster by these
experiments. Has conjecture and misrepresentation and pandering to
sensationalism become an accepted standard for *SCIENCE* reporting?[8]

In due course the MIT Committee identified early February 1975 as an
auspicious time for an international meeting on recombinant DNA. Berg
again suggested Asilomar as the venue. This meeting is thus often referred
to as Asilomar II, to distinguish it from the earlier far less contentious
Asilomar I gathering in January 1973.

These events could not have been more ill-timed from Stanley Cohen's
perspective. There is much at stake in the race to cross the finish line first,
and cutting-edge science has always been highly competitive — often
contentious, the more so the more cutting the edge. Hence, the issue of
sharing newly discovered "precious" reagents can and does arouse anger in
some. The unwritten "laws" of scientific collegiality generally acknowl-
edge the prerogative of investigators to defer sharing reagents until their
availability is announced in recognized scientific periodicals. But some
investigators forgo such niceties in the altruistic interest of the progress of
science — and of being collegial. Colleagues working in the same institution, especially those with whom the investigators have more than a casual
acquaintance are typically expected to be exempt from such embargos.

To his credit, Cohen dutifully complied with the majority of requests
for plasmid pSC101. But in so doing he stipulated that the vector could
not be used for constructing antibiotic resistance combinations that did
not already exist in *E. coli*. "I was mindful of the fact that genes encoding
resistance to penicillins were already present in *E. coli*, so I was leery
of anyone introducing new resistance capabilities," Cohen related. "But
I also knew of some resistance traits that were *not* expressed in *E. coli* and
the potential for creating new combinations of resistance genes was a
matter of concern to me."[9]

One of the earliest requests to Cohen (perhaps *the* earliest) for plasmid pSC101 came from his Stanford colleague David Hogness, who, as mentioned in an earlier chapter, had generously provided Cohen with workspace in his laboratory until Cohen was able to occupy a renovated laboratory in the Department of Medicine. In light of the stunning progress toward gene cloning, Hogness was eager to deploy the new technology for studies with Drosophila DNA.

Cohen found himself in a bind. He was unhesitatingly grateful for the courtesy he had been extended by Hogness and was initially inclined to donate the plasmid to his biochemistry colleague. But the recent work with Xenopus DNA had just begun, and he and Boyer (with whom he discussed the dilemma) were concerned about losing priority. "At that time we hadn't even produced an outline for a paper on the *Xenopus* work," Cohen told interviewer Sally Hughes in 2009. "We were still relatively early in the Staphylococcal work and were still a few months away from publication of our *E. coli* plasmid DNA cloning experiments. When I told Herb Boyer about Dave Hogness' request he suggested that I was crazy to even think about giving out anything at that time. After several restless days and nights I met with Hogness and told him:

David Hogness.
(http://hogness.stanford.edu/)

I have a lot of personal torment about this because I appreciate the help that I've received from you. Of course you can have the plasmid the day that our first paper about its use for DNA cloning is published. But, the DNA cloning work is probably going to be the most important research that I will ever do. I'm a junior scientist working in the department of medicine with clinical responsibilities and you're an internationally known molecular biologist. If I give you the plasmid and you publish the cloning of *Drosophila* DNA about the same time as we publish our work, the cloning of eukaryotic DNA will be seen largely as your discovery. I'd like to wait a few months before giving you the plasmid.[10]

Cohen's wish to unambiguously stake his claim in the highly competitive world of molecular biology and genetics in the mid-1970s was not unreasonable, and it is somewhat surprising (at least to this author) that someone of the stature of David Hogness would not have appreciated this nuance. One might surmise that there must surely have been a solution to this gridlock that would have satisfied both parties' concerns — short of waiting for Cohen's paper to be published. But Hogness was offended by Cohen's response and he petulantly and without permission retrieved some of the plasmid DNA that John Morrow had stored in his freezer in the Department of Biochemistry in anticipation of collaborating with Cohen and Boyer.

Hogness may well have reasoned that since Morrow was in fact a student in the Department of Biochemistry, or most certainly had been for an extended period, his (Hogness') actions should not be construed as being deliberately provocative. As word of this incident spread through the department, it reflected poorly on Hogness judgment. He later apologized for the episode. But when the full details of the occurrence (including Cohen's refusal to accede to Hogness' request) reached Berg's ear (he was then the department chair), he too was less than totally sympathetic about Cohen's decision.

"Paul was very angry about my decision to delay providing pSC101 to Dave," Cohen related. "He said that he and others in the department had been instrumental in bringing me to Stanford ⋯⋯⋯⋯ and that my refusal to give Dave the plasmid was so unreasonable that it had led him to do a foolish thing. Paul was really quite vituperative — I guess that is the word — and almost vengeful in his attitude, and I agreed to think further about my decision. ⋯⋯⋯⋯⋯⋯ That was a negative turning point in my relationship with the Department of Biochemistry. There are people in that department who frankly have never forgiven me."[11]

As already stated, Berg and Cohen did not then enjoy a trouble-free relationship and tensions between the pair about sharing biological reagents resurfaced some years later. In the early 1980s, when gene cloning was going full bore, Berg's laboratory was in search of mammalian genes that might serve as selectable genetic markers in the course of screening cells containing recombinant DNA molecules. In October 1978, Cohen and Robert (Bob) Schimke (a highly regarded investigator in the biology department at Stanford) and their colleagues published the results of a collaborative study on cloning the mouse dihydrofolate reductase (*DHFR*) gene.[12]

Intent on acquiring a cDNA for this gene for unrelated experiments, Berg asked Schimke (a colleague and friend with whom he frequently played tennis) to acquire some of the cloned cDNA. Since his and Cohen's work was published, Schimke foresaw no problem in providing it to Berg. But when a student from Berg's laboratory went to Schimke's laboratory to obtain the cDNA, he was informed that the material belonged to Cohen and he could not provide it to Berg's group. Berg repeatedly asked Schimke to forward his request to Cohen directly, and Schimke in turn repeatedly assured Berg that Cohen had no problem with giving him the reagent. But the wanted DNA never materialized!

"This went on for about a year," Berg related, "until I emphasized to Schimke that since the work from which this reagent was derived was published, this behavior was a gross breach of scientific custom — and that I was shocked that he would tolerate such an attitude by Cohen."[13] After some harsh words were exchanged, Berg wrote to Schimke:

> Losing my temper when we spoke the other night was a foolish and useless response. I'm sorry it came to that; but let me reiterate that I meant every word I said, it's only the tone I regret.[14]

Berg eventually obtained the reagent he sought from Schimke. "Sometime after the *sturm* and *drang* was over Stan showed up in my office with an offer to collaborate on experiments involving the *DHFR* gene," Berg related. "I asked him to leave my office at once!"[15] (Berg used a far less polite phrase when he related this anecdote to the author.)

As pointed out earlier, while Paul Berg can (and does) exhibit heights of anger on occasion, he is not one to hold grudges — and his temper cools almost as quickly as it erupts. At the time of this writing, the two pioneers of recombinant DNA and gene cloning enjoy a mutually respectful working relationship.

****

To conclude the momentous events of 1974, let's return to the weeks immediately following the MIT Committee meeting when the paper by Morrow *et al.* announcing the successful propagation of Xenopus DNA in *E. coli* appeared in print. The final paragraph of this paper left little room for any doubt that the era of gene cloning had arrived:

> The procedure reported here offers a general approach utilizing bacterial plasmids *for the cloning of DNA molecules from various sources* [author's italics], provided that both molecular species have cohesive termini made by a restriction endonuclease, and that insertion of a DNA segment at the cleavage site does not interfere with expression of genes essential for its replication and selection.[16]

The only mention by Morrow *et al.* of the paper by Mertz and Davis (published in the *PNAS* in 1972) in which the pair explicitly hypothesized the potential for gene cloning by joining SV40 and λ*dvgal* DNA, is a rather oblique sentence in the introduction that simultaneously acknowledges the Italian Vittorio Sgaramella in Josh Lederberg's laboratory.

> Recombinant DNA molecules constructed *in vitro* from separate plasmids by joining of fragments having cohesive termini (Mertz and Davis, Sgaramella) generated by the *Eco*R1 restriction endonuclease ....[17]

****

Prior to public release of the "Berg letter," Howard Lewis, the public information officer at the National Academy of Sciences had unilaterally arranged a press conference in Washington, DC. "To Lewis, academy committee reports and press conferences elucidating them were as inevitable as cause and effect. The Berg committee's open letter clearly constituted a committee report, and a press conference was taken for granted," John Lear wrote.[19]

With so many fingers in the proverbial pie misunderstandings were inevitable. Unaware of the academy's intentions to hold a press conference, the MIT Committee had become similarly proactive on its own recognizance. Indeed, committee member and MIT denizen David Baltimore had arranged for an exclusive press release by Victor McElheny, a science writer with the *New York Times* who had somehow gained access to the "Berg letter," though no specific time for this "exclusive" had been established.

Further muddying the broiling waters, Spyros Andreopolis, the public information officer at Stanford Medical School, also heard about the letter through the grapevine. Understandably chagrined that his office had been left out of the loop, Andreopolis contacted Berg requesting a copy of the letter, with the explicit intention of composing a press release from Stanford University. He dutifully shared a draft of the release with Berg, who immediately objected in light of the promise made to McElheny. But Andreopolis too was operating under a sense of obligation. As Stanford's public information officer he felt duty bound to notify all West Coast newspapers, especially in light of the obvious import of the news and Stanford's central involvement. He was unmoved by Berg's objections!

Ultimately the interested parties managed to get on the same page, though the solution that emerged did not please all concerned — in particular Victor McElheny. The *New York Times* released a story on July 18, 1974 (under the blazing headline "Genetic Tests Renounced Over Possible Hazards") the day that the national academy held its promised news conference. The story was clearly no longer a McElheny exclusive! Additionally, well experienced in journalistic know-how, a reporter for the *Washington Post* who had been alerted to the academy press conference smelled blood and used his journalistic skills to ferret out the story, which was published on the same day.[20] David Perlman, an experienced and adroit reporter for the *San Francisco Chronicle*, who over the course of the next several months and years wrote a number of engaging and notably accurate accounts on the recombinant DNA saga, was also not asleep on the job and independently provided his own story with the header: "A Danger in Man-Made Bacteria."[21] Naturally all these newspaper accounts totally eclipsed the *National Academy* press conference — which was held later that morning.

Following the press conference, Berg, Baltimore and Lewis retired to a luncheon meeting at which they agreed to reconvene the MIT Committee one more time to begin planning Asilomar II (officially designated as the *International Conference on Recombinant DNA Molecules*), a conference that turned out to be one of the most momentous and far-reaching in the history of science.

The MIT Committee reconvened on September 10, 1974. The meeting now included Donald Brown, Richard Novick and Aaron Shatkin, who had been summoned because they agreed to play key roles as chairmen of various working groups that would be established to issue formal reports at the meeting, Also present then was William (Bill) Gartland, who, in response to a forward-looking request from the MIT committee a few months earlier, was appointed by the NIH as secretary of an as yet unfilled committee dubbed the *Recombinant DNA Advisory Committee to the NIH Director* [Donald (Don) Frederickson]. The following month, the NIH formally sanctioned the establishment of a *Recombinant DNA Molecule Program Advisory Committee*, which was duly represented at Asilomar II.

Sensing that its work was essentially done, the MIT Committee was dissolved and replaced by a smaller more focused *Asilomar Organizing Committee* to deal with the nuts and bolts (of which there were a great many) of arranging an international conference just eight months hence. David Baltimore, Maxine Singer and Norton Zinder agreed to assist Berg in this capacity. Participation by scientists outside the US was addressed by extending invitations to join the organizing committee to Niels Jerne, the highly regarded Director of the Basel Institute for Immunology, and to Sydney Brenner of the Medical Research Council Laboratory of Molecular

Biology in Cambridge, UK, and a leading light in almost anything to do with biology — and more besides! The MIT group also identified a cadre of prominent molecular biologists, microbial geneticists, virologists and relevant others to organize and chair the small science panels mentioned above, which would be responsible for formulating and initiating discussions of key topics at Asilomar. In the final analysis panel presentations were organized and/or presented by Stanley Falkow, Richard Novick, Aaron Shatkin and Donald Brown. These pre-Asilomar deliberations of these panels proved to be of inestimable value when the meeting finally got underway.

Notes

1. Personal communication from Paul Berg, October 2012.
2. PB interview by Rae Goodell, MIT Oral History Project, May 1975.
3. Letter from Paul Berg to Patricia Berg, February 11, 1974.
4. Berg P, *et al.* (1974) Potential Biohazards of Recombinant DNA Molecules. *SCIENCE* **185**: 303.
5. ECF interview with Paul Berg, October 2012.
6. Wade N. (1974) Genetic Manipulation: Temporary Embargo Proposed on Research. *SCIENCE* **185**: 332–334.
7. *Ibid.*
8. Letter to Phillip Abelson, February 24, 1977. Reproduced with permission from the Department of Special Collections, Stanford University Libraries.
9. SC interview by Sally Hughes, Regional Oral History Office of the Bancroft Library at the University of California, Berkeley, 1995.
10. *Ibid.*
11. *Ibid.*
12. Chang ACY, Nunberg JH, Kaufman RJ, *et al.* (1978) Phenotypic expression in *E. coli* of a DNA sequence coding for mouse dihydrofolate reductase. *Nature* **275**: 617–624.
13. ECF interview with Paul Berg, September 2011.
14. Note from Paul Berg to Robert Schimke, dated February 5, 1979. Reproduced with permission from the Special Collections, Stanford University Libraries.
15. ECF interview with Paul Berg, September 2011.
16. Morrow JF, Cohen SN, Chang ACY, *et al.* (1974) Replication and Transcription of Eukaryotic DNA in *Escherichia coli*. *PNAS* **71**: 1743–1747.
17. *Ibid.*

18. Falkow S. (2001) I'll Have the Chopped Liver Please, or How I Learned To Love the Clone: A Recollection of Some of the Events Surrounding One of the Pivotal Experiments That Opened the Era of DNA. *ASM News* **67**: 555–559.
19. Lear J. (1978) *Recombinant DNA: The Untold Story.* Crown Publishers Inc., p. 90.
20. *Ibid.* p. 91.
21. *Ibid.* p. 91.

CHAPTER TWENTY

Asilomar II

An outright ban would be a disaster

Niels Jerne declined the invitation to join the Asilomar Organizing Committee, but Brenner accepted. A legendary "outside the box" thinker, Brenner had already been involved in a similar strategic review and planning effort in the UK, where little time had been lost in coming to grips with the challenges and conundrums presented by the new DNA technology. A Working Party on the Practice of Genetic Manipulation was assembled under the watchful eye of Lord Eric Ashby, a distinguished British plant physiologist and Master of Clare College at Cambridge University, and a figure highly regarded for his grasp of scientific issues that spilled over to society at large. Brenner was invited to appear before the Working Party as an expert witness. He (predictably) did outstanding homework and delivered a carefully considered White Paper with the following opening paragraph:

> The recent development of biochemical methods for joining DNA molecules together with their subsequent introduction into bacteria or animal cells has produced a qualitative change in the field. It cannot be argued that this is simply another, perhaps easier, way to do what we have been doing for a long time with less direct methods. For the first time, there is now available a method which allows us to cross very large evolutionary barriers and to move genes between organisms which have never had genetic contact.[1]

Perhaps Brenner's most notable contribution to the Ashby Working Group (besides his omnipresent and irrepressible humor) was a suggestion that became pivotal to the subsequent Asilomar audience.

When we come to consider the hazards of this kind of work we enter into a very difficult area. .... It is therefore necessary to consider in detail how this work might be controlled. An outright ban would be a disaster, since it would deny us access to the very scientific knowledge we urgently need. Many microbiologists have pointed out that there are containment procedures that can be applied to lessen the dangers. [The reader may recall an essentially identical suggestion from Janet Mertz when she attended a workshop at the Cold Spring Harbor course she attended in the summer of 1971 (see Chapter 16).] In addition, I believe that much can be done and should be done by molecular biologists to engineer the microorganisms used so that their chances of survival outside a laboratory environment are small. By multiplying a set of small probabilities we could reduce the hazard to very small numbers, say 10^{-24} (reciprocal of Avogadro's number) or even 10^{-50} (about the reciprocal of the number of elementary particles in the universe).[2]

The Ashby Working Party submitted its formal report in January 1975. The document included the following statement:

Finally, we would put on record our impression of the general concern among scientists themselves over this matter. It was a group of research workers who called for a pause in the work to give time for the risks and benefits to be assessed. It can be no more than a pause, because the techniques open up exciting prospects both for science and for its applications to society, and the evidence we have received, indicates that the potential hazards can be kept under control.[3]

Despite its measured tone, the Ashby report inevitably fell short of the desires of some publicity-hungry critics, forewarning media reaction still to come. Bernard Dixon, writing in the *New Scientist* on January 23, 1975, commented:

.... the report falls down badly. .... while suggesting that "this search for genetic safeguards should be pursued," the report concludes that genetic manipulation should continue to be used. No embargo. Not even a voluntary pause while such practical safeguards are developed.

The Ashby group specifically points out that it was not concerned with wider ethical questions about the use of the new techniques. Even so, it is surely unsatisfactory for the working party to have reported for the first time the persistence (even temporarily) of *E. coli* K12 in the gut; to suggest model systems that might be developed to obviate all risk; and yet to conclude that not even a brief moratorium is needed. That it should have reached this disquieting conclusion, just a month before an international conference in California is to consider the whole question, is even more regrettable.[4]

Having received a copy of the Ashby Working Group report in advance of the Asilomar II meeting, Berg pressed hard to acquire Brenner's participation as a member of the Organizing Committee; perhaps "the wisest decision concerning the entire conference," in his (Berg's) retrospective view.

Brenner gave an interview for the MIT Oral Program in May 1975, a mere few months after Asilomar II and hence at a time when his recollections were presumably acute and largely if not entirely accurate. Not surprisingly, the transcript is engaging, if only to acquire a sense of Brenner's legendary sense of humor and his distinctive approach to scientific issues. For example, when asked for his views on growing perceptions

The ever-impish Sydney Brenner.
(Courtesy of Paul Berg.)

of a more relaxed attitude to warnings of a moratorium on recombinant DNA research in Europe, he told interviewer Charles Weiner:

> You have to be careful to distinguish between impotence and chastity [a distinction between not *able* to do something (impotence) and not *wanting* to do something in this operation (chastity)]. Most people in Europe just weren't *able* to do this work; they simply didn't have the technology. Therefore chaste was very easy. And I think that a lot of the apparent chastity displayed by Europeans' restraint was that — essentially impotence.[5]

With February 1975 looming ominously on the horizon, Berg and his colleagues had their work cut out for them. In those pre-email days, communication relied on the telephone and conventional "snail" mail. In early December 1974, a formal letter of invitation was drafted and mailed to about 150 invitees whose expressed opinions might be of value.

> Dear ········,
>   A meeting on Recombinant DNA Molecules will be held in Pacific Grove. California (at the Asilomar Conference Center) during February 24–27, 1975. ········· The purpose of the meeting is to review the progress, opportunities, potential dangers and possible remedies associated with the construction, and introduction of new recombinant DNA molecules into living cells.········· The program will include discussions of (a) the present and future methodologies for constructing, propagating and amplifying recombinant nucleic acid molecules; (b) the molecular biology and natural history of autonomously replicating plasmids and their microbial plant and animal hosts; (c) an examination of the rationale and potential risks of introducing such genetic infection into microbial plasmids; (d) the methodology, scientific and practical benefits, and the risks of linking segments of eukaryote genomes to autonomously replicating microbial plasmids; (e) approaches to assess and to minimize or eliminate any serious biohazards stemming from this line of research.[6]

Berg solicited suggestions for potential participants. But the final invitation list was of his selection. About 90 of the invited scientists were American and another 60 came from 12 different countries. All the scientists were among the elite in one segment of molecular biology or another. No organizations were represented per se. Sixteen members of the press were also invited, all accepting the condition that copy would not be filed until the three-and-one-half-day conference had ended.

Following considerable discussion with the chairs of each of the panels that convened in anticipation of the meeting, the following primary topics emerged in the final conference program: *The Ecology of Plasmids and Enteric Organisms* (chaired by Richard Novick), *The Molecular Biology of Bacterial Plasmids* (chaired by Aaron Shatkin) and *Plasmid-Viral DNA Recombinants* and *Plasmid-Cell Recombinants* (chaired by Don Brown), each with a slate of expert speakers.

"The working groups made a huge difference," Maxine Singer related. "We met with them on an ongoing basis and it was evident that they were writing productive and thoughtful reports."[7]

The final evening of the three-day meeting was reserved for presentation of ethical and legal issues, and the final morning was reserved for open discussion and adoption of (an anticipated) consensus conference statement.

Berg (right) conferring with British scientist Walter Bodmer. (Courtesy of Paul Berg.)

A comprehensive recounting of the many twists and turns that visited Asilomar II is beyond the scope of this book. Those interested in reading more complete renderings of the meeting are advised to consult the numerous published contributions on this topic listed in the reference section. But, as Jim Watson and John Tooze discerningly commented in their 1981 literary contribution *The DNA Story — A Documentary History of Gene Cloning*, it would be extremely difficult to improve upon Michael Rogers' masterly description of the atmosphere at the Asilomar Conference.[8]

This evocative piece by Rogers was published in *Rolling Stone* magazine on June 19, 1975, just four months after the meeting. It bears the expressive title: *The Pandora's Box Congress — 140 Scientists Ask: Now That We Can Rewrite the Genetic Code, What Are We Going to Say?*[9] The article preceded Rogers' more extended and better-known account in a book called *Biohazard*, published in 1977.[10] In light of the historical significance of Asilomar II, parts of Rogers' facetious, amusing and atmospheric June 1975 *Rolling Stone* article are reproduced here verbatim. Other salutary accounts of the conference are to be found in most of the many books published about the recombinant DNA controversy (see references). Rogers wrote:

> The conference — four intense 12-hour days of deliberation on the ethics of genetic manipulation — should survive in text yet to be written, as both landmark and watershed in the evolution of social conscience in the scientific community. ·············· The molecular biologists descend upon California's Monterey Peninsula on very

Michael Rogers from *Rolling Stone*.
(From BIOHAZARD, Deacon Chapin with Rolling Stone magazine, 1977.)

nearly the same day as do the monarch butterflies ·········· numbering in the millions. ··········· They arrive on a bright blue Sunday afternoon — the finest February weather that the Monterey Peninsula has to offer — shuttled in from the local airport, often still clad in overcoats donned earlier in Cambridge or Krakow.

First Session

Monday morning, ··········· the molecular biologists begin to file ············ into the redwood chapel that will serve as center for the next four days. ·········· Debating the ethics of human interference with the mechanics of evolution in a church at the edge of the immense saline test tube where it all started. Rarely does one find one's metaphors so cheap — or so apt. "Here we are," a young scientist from the East Coast will tell me later that night over beers, "sitting in a chapel, next to the ocean, huddled around a forbidden tree, trying to create some new Commandments — and there's no goddamn Moses in sight." ························

Paul Berg, the Stanford biologist who headed the moratorium group just doesn't look enough like Charlton Heston. This morning he appears the quintessential young California academic: tanned, intense, athletic, in studiedly casual sport clothes and a suitably collegiate sweater donned against the early morning Monterey chill. He might as easily be dressed for sailing or an early round of golf, but here he is, standing onstage in front of 140 international colleagues, expressing once again a concern that some privately consider obsessive: "What is new," he says with flat certainty, "is that recombinant DNA can now be made from organisms not usually joined by mating — and hence can give rise to DNA molecules not previously seen in nature."

"If we come out of here split and unhappy," says another conference organizer — a young, successful molecular biologist named David Baltimore, clad in trim beard and elaborately embroidered Levi Jacket — "'then we will have failed the mission before us."

The first session rolls on well past the lunch bell, much of it fetchingly anal. A major question, vociferously argued, is just how likely it is that these *E. coli* K12 bacteria, so long laboratory pampered, will survive in the human gut should they escape their test tubes. A series of British researchers demonstrate a consistent penchant for mixing K12 cultures into half pints of milk, swallowing same and then monitoring their subsequent stool for evidence of bacterial survival. The topic offers some opportunity for drollery ········ but by the end of the session, the implications of K12 ingestion seem far from resolved. But by the end of the same session, another implication seems all too clearly defined: pure and unadulterated paranoia.

The Press, The Public and Paranola

An embargo was a key feature of press coverage of [this] historic scientific conference. ······························ From the outset, planners envisioned inviting eight reporters to cover the meeting, on the condition that they would agree not to file any reports on the meeting until after its conclusion. ···················· However, faced with a suit by the *Washington Post* for access to cover the meeting, Howard Lewis the academy's press officer, convinced the organizers to expand the reporters' quota, first to twelve spaces and eventually to sixteen.[10A]

"These proceedings," announces David Baltimore ············, "will be taped — for the archives and for review, not for release. And anyone who does not want to be taped may ask and the machine will be turned off." Immediately someone rises in the audience. "But what about the *press?*" There is a brief silence in the still somnolent audience. What *about* the press, with those nasty Sony cassette machines perched stage left, right beside the official Academy of Science sound equipment? After some deft reassurance by the NAS press officer, a vote is taken. The press is permitted, with many abstentions, its recording equipment. But it is not, by any means yet permitted any real welcome.

And that's no surprise to the press. By now the vibes are unmistakable — and have been so since first application for permission to attend Asilomar. Press attendance was not actively encouraged by anyone involved and in the case, say of a reporter from *ROLLING STONE*, it took some persistence even to find out whom to *ask*. A writer from Washington told the conference organizers straight out: "A secret international meeting of molecular biologists to discuss biohazards? If the press isn't allowed, I'll guarantee you nightmare stories." Or as a journalist from Southern California said: "The scientists loved the press when we got Nixon. But when we start hanging around their own back yard, they act very nervous."

It seemed almost a paradigm of the unsatisfactory relationship between the press and science, paranoid behavior is guaranteed to engender the journalistic suspicion that something is up. And the more attention the press paid, the more paranoid the attendees became and not entirely without reason. A suitably hysteric story about the antics of an international cabal of biologists devoted to some blackly humorous campaign of rearing new cancer viruses might be just the thing to stampede Joe Public.

Welcome or not, however, the press was there, hunkered down in the front row, Sonys turning, and there was really nothing to be done about it. By the end of the sessions, it was clear that the press presence caused the conference attendees both some discomfort and some extra efforts toward public caution. And that, finally, seemed fairly healthy.

Happy Talk

During the first two days of sessions, it becomes immediately clear that the conference would really rather talk about almost anything than the issue at hand. ⋯⋯⋯⋯⋯⋯ The talk is exceedingly technical and wanders over a spectrum of topics — some including information sufficiently original that afterward researchers queue up at Asilomar's two pay telephones to relay the word back to their laboratories. ⋯⋯⋯⋯⋯⋯⋯⋯⋯

[E]ach time the real issue arises — as, say, when some Advisory Group, previously appointed, introduces *"Proposed Guidelines for Plasmid-Cell DNA Recombinant Experiments"*, the proceedings rapidly develop the appearance of some obscure primitive tribe eons ago, accidentally stumbling by trial and error onto the secret of parliamentary procedure. Odd, one might think, that a roomful of the leading minds on the leading edge of science can't agree on how to run a meeting. But there is little evidence of any concerted drive toward organization here.

Sometimes the derailments are benign: An irrepressible gentleman from Switzerland monopolizes a microphone for a baffling ten-minute dissertation on scientific ethics, by the end of which it is difficult to decide whether the researcher's English or his thought processes are the more twisted. Or some well-intentioned American drones for an equivalent period about his role in the licensing of an obscure vaccine some years earlier, which, it grows quickly clear, has very little to do with

anything. Or the entire discussion lurches off into a swamp of technical detail: "Why can I use *Xenopus* DNA under low risk conditions, when Josh Lederberg has to use *Bacillus subtilis* under moderate risk?"
....

The reader may have (accurately) gleaned from the above passages that notwithstanding Michael Rogers' puckish sarcasm, progress in the first few days of the Asilomar Conference was less than spectacular. But there is widespread agreement that Sydney Brenner ultimately salvaged the meeting. In 2002, on the occasion of Brenner's 75th birthday, Berg wrote the following tribute to Brenner's role at Asilomar:

> To ensure coordination and consistency between the Ashby working party's deliberations and the Asilomar conference's discussions Sydney was invited to join the organizing committee's planning efforts. It was perhaps the committee's wisest decision. Throughout the conference Sydney's vision of the new technology's opportunities energized the discussions and helped lay the foundation for the conclusion that a way had to be found to proceed with the research safely. One of the conference's fundamental decisions, in part driven by his vision, was to assign a best estimate of potential risks to contemplated experiments and to devise guidelines for increasingly stringent requirements for minimizing these risks.
>
> Two ways were conceived that could achieve this end. One, physical containment, pertained to the nature of the facilities to be employed for particular experiments; --------. The other proposed barrier to inadvertent release of engineered organisms reflected Sydney's ingenious suggestion of biological containment: experiments that were judged to present the greatest possible risks were required to use as their hosts for cloning experiments, genetically modified bacteria that could not survive escape from the laboratory. [The reader might recall a similar suggestion three years earlier by Berg and his graduate student Janet Mertz]. Sydney's forceful personality and prodigious skill in microbial genetics helped convince the assembled scientists that "safe" plasmids, phages, and cells could be developed.

"Working with Sydney throughout the recombinant DNA debate, especially during the Asilomar Conference, was one of the more

Sydney Brenner (standing, left) and Richard Roblin peer over the shoulder of David Baltimore. (Courtesy of Paul Berg.)

rewarding aspects of the experience," Berg wrote. "His unique brand of humor kept us going; his creative way of attacking scientific issues inspired us all. And his deeply felt ethical responsibility motivated many to forgo some of their more selfish instincts."[11] Michael Rogers was certainly among the more effusive Brenner supporters. In his preemptory "Pandora's Box" article, he wrote:

> Brenner, a compact Englishman in his 40s, with bushy eyebrows, gleaming eyes and nonstop animation that blend to an impression midway between Leprechaun and gnome, soon emerges as the single most forceful presence at Asilomar. Repeatedly, when the sessions wander off into a technical morass that threatens to engulf larger considerations, Brenner rises to redirect deftly.
>
> "Does anyone in the audience believe," he asks, in one such redirection, "that this work — in prokaryotes at least — can be done

with absolutely no hazard?" There is no immediate response. "This is not a conference," Brenner goes on, "to decide what's to be done in America next week. If anyone thinks so, then this conference has not served its purpose.

"In some countries," he says, "this would be done by the government, and once guidelines were set and you broke them there would be no question of peer censure — the police would simply come out and arrest you." This is an opportunity, Brenner concludes, for scientists to show that they can regulate themselves — "to reject the attitude that we'll go along and pretend there's a biohazard and hope we can arrive at a compromise that won't affect my own small area, and I can get my tenure and grants and be appointed to the National Academy and all the other things that scientists seem to be interested in."

Brenner — an authority in the field — also took the lead in a series of afternoon sessions devoted to a task called "disarming the bug." The sessions are lively and well attended and they represent a curious tangent of mankind's new involvement in the processes of evolution. These people are trying to create novel organisms that are by design incapable of living in the real world.

"The competitive nature of the institutions themselves can affect the situation," he says one afternoon on the chapel steps. "Sooner or later, at some place like MIT or Stanford, some laboratory assistant may well contract something like leukemia — and he will sue the place for everything they've got."

At the time, it seems a fairly pessimistic scenario but by the time Asilomar lurches to its conclusion, it grows clear that Brenner has described an issue that ultimately proves thoroughly telling.[12]

Rogers' diatribe encompassed his views on the contributions of the two Nobel laureates at the conference — Jim Watson and Josh Lederberg, whom he facetiously dubbed "*Chosen Sons of Alfred Nobel.*"

Chosen Sons of Alfred Nobel

The two American Nobel Laureates at Asilomar — Joshua Lederberg from the West and James Watson from the East — exert powerful presences during the conference proceedings. And not always in a terribly popular fashion. "If we can't communicate the tentativeness of this document," Lederberg says early one afternoon ········ we are in trouble. "There is, he suggests ominously, "a graver likelihood of this paper crystallizing into legislation than some of us would like to think." ······················· Lederberg fears that such detailed restrictions might be taken too literally and inflexibly by some well-intentioned legislative body and thus thoroughly garrote future research.

Just before the dinner bell, Berg rises. "If our recommendations," he says, "look self-serving, we will run the risk of having standards imposed. We must start high and work down. We can't say that 150 scientists spent four days at Asilomar and all of them agreed that there was a hazard — and they still couldn't come up with a single suggestion. That's telling the government to do it for us."

The inspiration to invite a select group of lawyers, including her husband Dan, to participate in the meeting came from Maxine Singer. He was joined by a number of colleagues with similar leanings. Collectively this legal gathering gave the remaining conferees a speedy wake-up call — deftly recounted by Rogers:

The Lawyers — Reality Therapy

By Wednesday night — with only a single morning session remaining — the situation at Asilomar seems as unsettled as the Monterey weather. The mild blue skies have begun to turn and by now a massive bank of thick gray fog lies a mile or so off the coast, sending in low, damp clouds both morning and evening.

The Wednesday night program looks fairly innocuous: presentations by lawyers regarding ethics and legal liability. ---------------- The second speaker — a professor of international law promises to be entertaining. He starts with a -- joke: A scientist and a Iawyer are arguing about which of theirs is the older profession. The argument goes back and forth, from Pericles to Hippocrates to Maimonides to Hammurabi, until it reaches all the way back to God. God, the scientist states, must have been a scientist to have brought order out of chaos. Yes, the lawyer responds. But where do you think the chaos came from?

The joke turns out to be less joke than promise, as the young lawyer proceeds into a merciless "outsider's analysis" that, within minutes, has jaws dropping all over the chapel. Much of the conference, he suggests, has been irrelevant to the central issue. ----------- Many of the specific arguments, he goes on, have been equally inapplicable. "Academic freedom," he points out, "does not include the freedom to do physical harm..." "This group," the young lawyer suggests flatly, "is not competent to assign overall risk. ---------------- It is the right of the public to act through the legislature and to make erroneous decisions."

But the best is yet to come. The third speaker [a lawyer named Harold Green] appears something of a manic milquetoast, but by the time he is behind the podium and has completed a sentence or so, his dry sharp delivery makes it clear that this is the lawyer who slices one mercilessly to very tiny ribbons in the witness box. Which is, coincidentally, precisely his topic: "Conventional aspects of the law," as he puts it, "and how they may sneak up on you — in the form, say, of a multimillion dollar lawsuit."

Professional negligence, the lawyer suggests, is a failing that finds juries exceedingly unsympathetic. Judges, he points out are expert only in the law — and juries, quite intentionally, are experts in nothing at all. By now, the gathered molecular biologists are conspicuously attentive.

Reports from David Perlman, a long-time journalist for the *San Francisco Chronicle,* were especially informative and salutary for his alarm-free tone and avoidance of gratuitous sensationalism and hyperbole in writing for a conversant lay public. On Saturday, February 22, 1975, Perlman set the stage for the meeting to begin two days hence:

The conference at Asilomar, near Monterey, will bring together more than 150 scientists from 16 nations ranging from the Soviet Union and Poland to Japan and Israel.

It is almost unprecedented in the history of science because the meeting's leaders hope the participants will reach agreement on a set of self-imposed restrictions on their fundamental research that will be honored throughout the world.

The idea of restricting research is a thorny one to scientists, yet there is broad agreement that some recent experiments leading toward the manipulation of genes in bacteria and other organisms may in fact carry serious but as yet unassessed dangers.

At the same time, the research offers the potential for incalculable human benefits. These could include new approaches to the treatment of many types of cancer and hereditary diseases; new methods for cheaply mass-producing vital hormones and antibiotics, and new bacterial substitutes for the fertilizers so badly needed by an increasingly food-short world.

The basis for such ambitious hopes lies in a field of genetics research that has burgeoned with unparalleled speed in only the past two or three years.[13]

On Friday, February 28, 1975 Perlman posted another low-key but informative overview:

— although their laboratories compete as bitterly as baseball teams for first place in the discovery race, they reached near-unanimous agreement on their general guidelines for public safety in their future experiments. They classified experiments according to their degree of potential hazard. They recommended rigorous, and in many cases highly expensive modification of laboratories in order to keep genetically altered experimental germs from escaping. And they set standards for matching the hazards and the safety measures according to a graded scale. They also urged biologists and geneticists in every country to develop within their own traditions the processes by which safety controls can be self-imposed or regulated without blocking further research.

The long and complex draft of the conference recommendations will now be formalized in greater detail by Berg and his colleagues, for official approval by America's *National Academy of Sciences.*

Beginning today in San Francisco, a special advisory committee of the National Institutes of Health will undertake the difficult job of drafting detailed safety standards that will officially regulate the experiments of American scientists working in this new frontier field of biology.[14]

Asilomar II took a huge physical and mental toll on the entire Organizing Committee — but on no one more so than Berg, whose attention and vigilance were constantly required. The committee convened every night at the conclusion of the evening session — typically late, to review and assess progress of the day's proceedings and to plan for the days ahead. "We were up until the wee hours on more than one occasion," Maxine Singer related.

"The lawyers were also very helpful," Singer continued. "One could not miss the change in attitude when Harold Green got up and told them what would happen if somebody filed a lawsuit! That changed things a lot." Singer parenthetically added: "After all is said and done Paul's chairmanship of that meeting was masterful. I know that I couldn't have done it."[15]

Harold Green, the lawyer whose warnings of possible expensive and debilitating legal consequences by way of suits filed by Joe Public was one of the attendees whom Berg took the trouble to thank in writing for their contributions to the meeting. In responding with a thank-you note, Green wrote:

> I am most appreciative of your kind words, and I would like you to know that meeting you, and watching you in action, was for me the high point of the entire exercise. As I observed to Dan and Maxine, I have rarely encountered anyone, especially a scientist, with the breadth of comprehension, understanding, diplomacy, and sheer competence that you exhibited.[16]

"The one thing the organizing committee had absolutely determined was that we were not leaving Asilomar until everyone had agreed to do *something* [author's italics]," Singer commented.[17] The night preceding the final day of the meeting was especially onerous since the Committee had resolved to present a draft report to the participants the following morning prior to closing the meeting. This effort didn't reach a satisfactory conclusion until the sun was about to rise. "When we were eventually done we were totally exhausted," Berg wrote. "But we were also so euphoric about having orchestrated a meaningful and constructive report that we took a long walk on the beach in the fading moonlight. We felt that we'd just written the Declaration of Independence!"[18]

Berg had no conscious intention of indulging prolonged debate on the content of the report — or even requesting its formal approval by vote. But Sydney Brenner wisely suggested that the committee consider asking for a straw vote approving a decision to the effect that recombinant DNA research should be resumed, but with appropriate safeguards.

As the bell began insistently announcing the final lunch of the meeting, Berg, making himself heard over the commotion, shouted: "All those in favor of this as a provisional statement, please raise your hands." A mere four or five (no one took an accurate count) hands were raised in response to the following call for negative votes.[20] The dissenters predictably included Josh Lederberg and Jim Watson, as well as Stan Cohen (who protested that he never voted on anything he hadn't read himself), Waclaw Szybalski (a scientifically respected but frequently provocative and contentious professor at the University of Wisconsin who we shall presently encounter more fully) and the Frenchman Philippe Kourilsky.[21]

> After we voted it was clear that a lot of people had simply sat and listened and this "silent majority" had made up their minds that there was indeed a reasonable middle path to follow. The edges were still very unclear, but there was a reasonable middle path to follow that they could see themselves implementing ------. When this was done I realized that there was no way to avoid central control of some type![22]

In the final analysis, certainly in the minds of the exhausted organizing committee, "the statement with which they had begun the morning — although frayed and variously patched along the way — had made it through, still holding to the framework fashioned by the organizing committee in their last night's vigil."[23]

A formal meeting report by Berg, Baltimore, Brenner, Roblin and Singer entitled "Summary Statement of the Asilomar Conference on Recombinant DNA Molecules" was published in the *PNAS* in June 1975.[24] A mere three and one half printed pages, the report is concise and to the point. The Introduction and General Conclusions read as follows:

> Of particular concern to the participants at the meeting was the issue of whether the pause in certain aspects of research in this area, called for by the *Committee on Recombinant DNA Molecules of the National Academy of Sciences*, U.S.A. in the letter published in July, 1974 should end; and, if so, how the scientific work could be undertaken with minimal risks to workers in laboratories, to the public at large, and to the animal and plant species sharing our ecosystems.
>
> The new techniques, which permit combination of genetic information from very different organisms, place us in an area of biology with many unknowns. Even in the present, more limited conduct of research in this field, the evaluation of potential biohazards has proved to be

extremely difficult. It is this ignorance that has compelled us to conclude that it would be wise to exercise considerable caution in performing this research. Nevertheless, the participants at the Conference agreed that most of the work on construction of recombinant DNA molecules should proceed provided that appropriate safeguards, principally biological and physical barriers adequate to contain the newly created organisms, are employed. Moreover, the standards of protection should be greater at the beginning and modified as improvements in the methodology occur and assessments of the risks change. Furthermore, it was agreed that there are certain experiments in which the potential risks are of such a serious nature that they ought not to be done with presently available containment facilities.

In the longer term, serious problems may arise in the large-scale application of this methodology in industry, medicine, and agriculture. But it was also recognized that future research and experience may show that many of the potential biohazards are less serious and/or less probable than we now suspect.

When the meeting participants vacated Asilomar and quiet and tranquility were restored to this beautiful California haven, Berg could discharge a deep sigh of relief. There is little doubt that when the first alarm bells began to ring in 1973 he had not the remotest notion of the many complexities that his decision to find a way of generating recombinant DNA in order to study how mammalian genes function would entail, nor of the administrative time and energy required. Now deeply into his retirement years, Berg is nonetheless gratified to discover that in recent times the Asilomar II meeting has been viewed as something of a paradigm for approaching other scientific problems with compelling sociopolitical implications, and that his advice and council as to how to structure and lead such conferences are eagerly sought.

Notes

1. Friedberg EC. (2010) *Sydney Brenner — A Biography*. Cold Spring Harbor Laboratory Press, p. 185.
2. *Ibid.*, p. 194.
3. Watson JD, Tooze J. (1981) *The DNA Story: A Documentary History of Gene Cloning*. W. H. Freeman and Company, p. 22.

4. Dixon B. (1975) Not Good Enough. *New Scientist*, January 23, p. 186.

5. Interview by Charles Weiner, MIT Oral History Program, May 1975.

6. Paul Berg, personal communication.

7. Interview with Maxine Singer, February 2012.

8. Watson JD, Tooze J. (1981) *The DNA Story: A Documentary History of Gene Cloning*. W. H. Freeman and Company, p. 25.

9. Rogers M. (1975) The Pandora's Box Congress. *Rolling Stone*, June 19, p. 37–42; p. 74; p. 77–78; p. 82.

10. Rogers M. (1977) *Biohazard: The Struggle to Control Recombinant DNA Experiments, the Most Promising (and Most Threatening) Scientific Research Ever Undertaken*. Alfred A. Knopf.

10A. Kierman V. (2006) *Rembargoed Science*. University of Illinois Press.

11. Berg P. (2002) Going Strong at 75: Asilomar and Recombinant DNA. *The Scientist* **16**: 16.

12. Rogers M. (1977) *Biohazard: The Struggle to Control Recombinant DNA Experiments, the Most Promising (and Most Threatening) Scientific Research Ever Undertaken*. Alfred A. Knopf.

13. *San Francisco Chronicle*, Saturday, February 22, 1975. Courtesy of David Perlman.

14. *San Francisco Chronicle*, Friday, February 28, 1975. Courtesy of David Perlman.

15. Interview with Maxine Singer, February 2012.

16. Letter from Harold Green to Paul Berg, April 7, 1975. Reproduced with permission from the Department of Special Collections, Stanford University Libraries.

17. Interview with Maxine Singer, February 2012.

18. Friedberg EC. (2010) *Sydney Brenner – A Biography*. Cold Spring Harbor Laboratory Press, p. 187.

19. Interview with Charles Weiner, MIT Oral. History Project, May 21, 1975.

20. Fredrickson DS. (2001) *The Recombinant DNA Controversy: A Memoir: Science, Politics and the Public Interest*, ASM Press, p. 25.

21. *Ibid.*

22. Interview with Charles Weiner, MIT Oral. History Project, May 21, 1975.

23. Fredrickson DS. (2001) *The Recombinant DNA Controversy: A Memoir: Science, Politics and the Public Interest*, ASM Press, p. 26.

24. Berg P, Baltimore D, Brenner S, *et al.* Summary Statement of the Asilomar Conference on Recombinant DNA Molecules. *PNAS* **72**: 1981–1984.

CHAPTER TWENTY ONE

The Dissenters: A Different Point of View

*All I can say today is I was a jackass
to sign the "Berg" letter*

In a piece published in *SCIENCE* in 1975, *New York Times* correspondent Nicholas Wade labeled Nobel Laureates Jim Watson and Josh Lederberg "a pair of *enfants terribles* [who were] constantly discovering holes in the committee's positions and ·········· breaking the ice for the faction among the younger scientists who were eager to get the moratorium lifted on the easiest terms feasible."[1]

Reluctantly or not, as a member of the MIT Committee Watson was ultimately a signatory of the written proclamation to convene an Asilomar II type of conference. But he is said to have subsequently referred to the meeting as "a waste of time," suggesting that the conferees "pretended to act responsibly, but actually were irresponsible in approving recommendations that did not adversely affect the work of anyone present." Later in the meeting, Watson was even more litigious:

> There is no logical way to prove that what we are doing is safe. All I can say today is I was a jackass to sign the "Berg" letter. It was not a well thought [out] document, but a silly emotional response which we scientists should all be ashamed of. I would willingly go to jail, say for thirty days; to atone for the harm I've caused others.[2]

Two years after Asilomar II, Watson offered additional explanations for his negative stance in an article entitled "An Imaginary Monster — The Only Danger We Face is the Specter of Untested Regulations," published in the *Bulletin of the Atomic Scientists*:

> My thinking at Asilomar was strongly influenced by the relative ease with which DNA molecules can pass into cells and be incorporated in a genetically active form into the chromosomes of the new hosts. When this phenomenon was discovered in 1943, it was thought to be restricted to the pneumococci bacteria. But over the years we have learned of more and more experimental cases where genetically active DNA has been transferred to a variety of different bacteria as well as to the cells of many higher organisms. That being the case, it seems probable that such transfers will also occasionally occur normally *in vivo*.
> ....
> Unfortunately, the majority of attendees at Asilomar were not prepared to say that the matter was hopelessly overblown. Instead they argued that even though no one was able to give any quantitative arguments as to dangers, it might be prudent to repress the most paranoid of their bad dreams by creating the so-called "safe *E. coli* K12 strains."
> .... The argument prevailed that even if we all believed that the ordinary laboratory variety of *E. coli* K12 was perfectly safe, why not give society another factor of safety?
> The trouble, however, with this way of thinking, as Joshua Lederberg was the first to point out, was that by creating the so-called "safe strains" the general public might come to believe that our ordinary laboratory strains of *E. coli* were, in fact, dangerous. ....
> .... In a real sense, by emphasizing the concept of the safe strain at Asilomar, the scientific community announced that it was worried that it might get the public in trouble. So the National Institutes of Health had no choice but to come up with regulatory guidelines. But given the absence of any firm facts implicating real dangers, the guidelines necessarily have to appear capricious, if not self-serving, in their details, and so are not easily defensible on any quantitative ground. So what started out as an attempt of the scientific community at appear responsible takes on increasingly the aspect of a black comedy.[3]

Sydney Brenner and Jim Watson in earnest discussion at Asilomar. (Courtesy of Paul Berg.)

By 1979, Watson's position was essentially unchanged, but he was more measured in his evaluation of the meeting and he emphasized a nuance not previously overtly considered: that molecular biologists (as opposed to other basic life scientists) were particularly vulnerable to recombinant DNA jitters! At a meeting on *Recombinant DNA and Genetic Experimentation* convened at Wye College, Kent, he stated:

> My thesis today is that the "Berg" letter was largely a reflection of the anxiety created within a number of molecular biologists about working on tumor viruses. This anxiety was felt most by people who had grown up in molecular biology as distinct from medicine or microbiology. Many of us had started working in science with *E. coli*, and subsequently went into tumor virology ⋯⋯⋯⋯⋯⋯. However, at the same time there might be biohazards created by this work. The facts on this latter point were scanty.
>
> When I became the Director of Cold Spring Harbor in 1969 and wanted to start up a tumor virus group, I, myself, had to face up to the question whether there might be any hazard to the scientists --- or to the community of families that lived very close to the labs. We, of course, suspected there was not any risk. ⋯⋯⋯⋯⋯⋯
>
> Now in looking back we have to ask the question of why we took so seriously Rube Goldberg schemes for getting cancer genes into us, when, in fact, we were already doing experiments which were infinitely more likely to put us at risk, if, in fact, any risk at all exists. The answer,

I believe, is that in running our labs we had already used up all our potential for worry. Any more accusations of potential moral indifference were more than we wanted to take psychologically.[4]

Josh Lederberg, the other Nobel laureate in attendance at Asilomar, was another consistent opponent of the recommendations that emerged from the meeting. He took the position that there was already too much regulation of research in the US and that those in support of recommendations from the Asilomar Organizing Committee "should carefully consider the consequences of stating such recommendations."[5]

Dissention about the merits (or lack thereof) of Asilomar II was totally predictable of course. As David Baltimore aptly stated in a letter to Berg some months after the meeting:

When we put the Asilomar meeting together we saw it as a group of responsible individuals trying to make a defensible stand on an issue with huge areas of total uncertainty. What we are seeing now is the people who do not feel the responsibility of maintaining the progress of science, and who are too damn certain about the uncertainties, taking over.[6]

Accusations of self-interest among molecular biologists also surfaced. Harvard biologist Richard Novick heroically addressed the touchy issue of self-interest — an oft-repeated theme during the meeting:

I would like to say, quite frankly, that my own personal interest in these experiments colors my view inevitably. In fact, given that I would like to do certain experiments involving R-DNA — experiments that I believe to be non-dangerous — I am quite unable to distinguish between two alternatives as the basis for this belief:

1. I am convinced they are not dangerous, and therefore it is okay to do them; or
2. I have convinced myself they are safe precisely because I want to do them!
   And I would defy *anyone* whose self-interest is involved in a particular line of activity to make that distinction.[7]

Predictably, as the person primarily responsible for initiating and organizing detailed consideration of the entire biohazard issue, Berg was a prominent target of accusations about his own self-interest — and that of Stanford University, especially when word leaked out that Stanford was applying for a patent on recombinant DNA (see Chapter 26). In early 1977, a French colleague wrote Berg a revealing note in this regard:

Dear Paul,

I would like to tell you ⋯⋯⋯ that two opinions have spread, among others, in Europe: (1) that Stanford asked for "safety limitations" to block others from working in genetic engineering: (2) that Stanford tries to keep for itself the possible economic advantages of scientific discoveries in a whole new field.[8]

Berg responded to these accusations with characteristic alacrity and passion:

It is an outright lie to imply ⋯⋯⋯ that the people who thought of and drafted the "moratorium" letter knew about the patent discussions or intention. ⋯⋯⋯ I and the other people who proposed and drafted the [Berg] letter were unaware of any patent application. I was the first of the group to learn of it ⋯⋯⋯ a week or two before the Asilomar meeting (Feb 1975). I objected to the application and have opposed it ever since. ⋯⋯⋯⋯⋯⋯⋯⋯⋯⋯⋯⋯⋯⋯⋯⋯⋯⋯⋯⋯⋯⋯⋯⋯⋯⋯⋯⋯⋯⋯

⋯⋯⋯⋯⋯⋯

My conscience is clear and my views about patents are on the record. I, personally, don't support the patent application, but I don't think it's illegal or immoral. I can understand why the French and possibly the Pasteur Institute might have their pride wounded and even feel that their opportunity to make money from their research in this field might be compromised. Very likely, if this work had been done at the Pasteur, there would have been a host of patents to ensure that its moneymaking abilities were ensured. It's easy for [Philippe] Kourilsky to be cynical and critical but it sounds like sour grapes to me; either that or some other Machiavellian reason.[9]

Sydney Brenner reserved his scorn for American scientists who seemingly opposed the Asilomar recommendation as a violation of academic freedom. In his May 1975 interview for the MIT Oral History Project, Brenner stated:

> Another thing that hit me at Asilomar, which I'd only had inklings of, is, I think, something about the social condition of American science. What bulked very large then, and I was astounded at, was this principle of academic freedom, that people said that *no one* has the right to tell me what experiments I may do and what experiments I may not do. ························ Someone stood up and essentially said: "It's a philosophical principle with me that no one can tell me what to do or not to do" ················· I was also struck by the tremendous fear of bureaucracy; this was very clear. And third, what I thought was uncanny, was what I would loosely call "right wing talk." People get up in America saying: "The Constitution gives the right to every American to bear arms and no one can tell me whether I can or cannot have a gun."[10]

As the meeting drew to a close, the organizing committee began to fret about gaining a consensus opinion from the participants. Once again Brenner came to the fore.

> About a dozen people did all the talking there. Lots of people talked in private; one got a sense of this. But most kept quiet in public. On the Thursday night [the last night of the meeting] the Organizing Committee was warned that there was obviously not going to be any agreement. We were at a loss and sat up the entire night writing the report. When we presented it to the meeting the next day, people asked how we were going to achieve consensus. --------------- I suddenly realized that we had to be sure that Asilomar was part of the mechanism of getting out of the moratorium. There was no way in which this could be maintained, in any form, and therefore it would be better to get out of this situation; not get out of the *principles* behind the moratorium, but out of the political situation, by getting a powerful consensus of opinion. That people were willing to show they were going to be restrained even about the way they thought about this issue. --------------------- So I said to Paul — we were both falling on our noses; we had an hour of sleep the night before — "take a vote on that principle." We did and there was overwhelming support. And after we voted it was clear that a lot of people had simply sat and listened and this

"silent majority" had made up their minds that there was indeed a reasonable middle path to follow. The edges were still very unclear, but there was a reasonable middle path to follow that they could see themselves implementing ------. When this was done I realized that there was no way to avoid central control of some type![11]

A note that Asilomar ended on is entertainingly captured in a clutch of letters that Berg received in 1977 from a group of fifth grade students who were studying genetics and had been told a little about the recombinant DNA controversy. "My students are extremely interested and have many questions, suggestions and opinions to offer you," their teacher wrote.[12] One of the student letters charmingly stated:

> Dear Sirs, I think you should continue your studies. But be careful about your chemicals so that it doesn't get out of the tube. Please send me the results. Keep up the good work guys and gals.[13]

A few weeks after returning to the blissful tranquility of his Stanford office, Berg penned a handwritten note to Maxine and Dan Singer:

> With a bit of time and rest I've acquired a new perspective on our Asilomar experience. Whatever happens, though, I shall never forget that incredible feeling of comradeship that developed that last night. No one who was not there will understand.[14]

Berg and Singer celebrated the 20th anniversary of Asilomar II with their own retrospective in the *PNAS*. Berg also revisited this history in 2000 (the 25th anniversary) and again in 2005 (the 30th anniversary). The 25th anniversary spawned a return to Asilomar by 67 scientists, lawyers, historians and ethicists for a two-day symposium, a meeting frequently referred to by Berg as Asilomar III. An opening keynote address by Don Frederickson was followed by commentaries by Berg, Robert (Bob) Sinsheimer, Maxine Singer, Stephen Stich and Hans Zachau. A special issue of *Perspectives in Biology and Medicine* published in 2001 features 10 articles that emerged from that meeting — including one by Berg, who posed the rhetorical question: If an Asilomar type of meeting took place in 2000, would it enjoy the same positive outcome?

In my view the Asilomar model would not succeed today to the extent it did 25 years ago for the following reasons. First, we should recall that the emergence of the recombinant DNA technology was sudden and unanticipated at the public level; the possibility that it carried potential dangers to public health removed it from the realm of just another interesting scientific advance. Furthermore, the news of this development came from the scientists conducting that research, not from some investigative reporter or disaffected scientist; that was most unusual, even historic. There seemed to be a need for consensus on how to proceed and a plausible plan for how to deal with this issue, both of which were provided by the scientific community. Action was prompt and seen by the public to have been achieved by transparent deliberations and with considerable impediments and cost to the scientists' own scientific interests. The issue and its resolution were complete before an entrenched, intransigent and chronic opposition developed. Attempts to prohibit the research or reverse the actions recommended by the conference were threatened, but such actions never generated sufficient reaction to succeed.

The issues that challenge us today are qualitatively different: they are often entwined with economic self-interest and increasingly beset by nearly irreconcilable ethical, religious, and legal conflicts, as well as by challenges to deeply held social values. An Asilomar-type conference trying to contend with such contentious views is, I believe, doomed to acrimony and policy stagnation, neither of which advances the cause of finding a solution. There are many forums for airing opposing views. But emerging with an agreed-upon solution from such exercises is elusive and discouraging.

The Asilomar II solution was arrived at by democratic means, and then an overwhelming majority voted in support of the recommendations advanced by the organizing committee. Lacking an overwhelming consensus in our society on the ethical issues concerning fetal tissue and embryonic stem cell research, genetic testing, somatic and germ-line gene therapy, and engineered plant and animal species, we are unlikely to move ahead except by political means wherein the majority prevails.[15]

These sentiments were reiterated in an essay by Berg published six years later as part of a series in *Nature* on "Meetings That Changed the

World." This particular article surfaced two further sage conclusions from Berg:

> In the 33 years since Asilomar, researchers around the world have carried out countless experiments with recombinant DNA without reported incident. Many of these experiments were inconceivable in 1975, yet as far as we know, none has been a hazard to public health. Moreover, the fear among scientists that artificially moving DNA among species would have profound effects on natural processes has substantially disappeared with the discovery that such exchanges occur in nature.
> ....
>
> That said, there is a lesson in Asilomar for all of science: the best way to respond to concerns created by emerging knowledge or early-stage technologies is for scientists from publicly funded institutions to find common cause with the wider public about the best way to regulate — as early as possible. Once scientists from corporations begin to dominate the research enterprise, it will simply be too late.[16]

Much has been written about the recombinant DNA controversy in general and Asilomar II in specific. Retrospective opinions may vary about the ultimate significance of that historic meeting, but only in degree. There can be little question that having been placed in the firing line from the very outset, Berg took the moral high ground at a time when others might have shrugged off the Bob Pollacks of the world in much the same way that Watson ultimately did.

Close to 40 years of incident-free experience tells us emphatically that there would have been no biological catastrophes, regardless of steps not taken and rules not promulgated. But, as repeatedly stated both in print and with the spoken word, the singular accomplishment for which Paul Berg's contributions must be acknowledged and celebrated is that in taking matters into its own hands at a very early stage, the community of scientists let its collective integrity and sense of responsibility be known to all. Little wonder that at various times during the decade 1970–1980, Berg was feted as a hero in various parts of the world. He also received a number of congratulatory letters from scientific colleagues who attended the meeting. His negative vote on the final morning of the meeting not-withstanding, the French scientist Philippe Kourilsky wrote:

> I would like first to congratulate you about the beautiful job which you made during the Asilomar Conference. Your efforts and courage were impressive and deserve the acknowledgments of all the scientific community.[17]

And in early March 1975, Robert (Bob) Sinsheimer (about whom more is written later), then chairman of the Division of Biology at Caltech wrote:

> I think you deserve many thanks and congratulations on your superb — even heroic — stewardship of last week's meeting. I think we reached a far more favorable outcome than I feared we might and it is certainly due largely to you and your organizing committee.[18]

Perhaps the most gratifying communication that Berg received immediately after the meeting was a handwritten note from his co-organizer David Baltimore — written while on a plane:

> If we were less inhibited people we would have had a good cry of exhilaration after the Asilomar meeting. Because we are not so demonstrative, I want to tell you before this ship lands and I resume my more usual style, how much I enjoyed working with you for the last week and how much I admire the even-handed but committed way you have guided this difficult but critical job to a successful and responsible conclusion. I know of no one else who could have done it so well — you continually surprise me with your insight and sensitivity. One of these days we should get drunk enough together to have the cry.[19]

The great majority of the Asilomar II conferees were pleased to be returning home after a week of intense and passionate debate. But the Asilomar organizing committee was not afforded this luxury. Instead they traveled directly to San Francisco to begin discussions with Hans Stetton of the NIH, who, together with a plethora of co-opted scientists would soon begin the arduous task of establishing federal guidelines for executing recombinant DNA research.

As stated in the preface, this period in the history of the early era of recombinant DNA is not examined in any detail in this book. However, the following three chapters are designed to give the reader a sense of the enormous scale of obligatory regulatory work that followed Asilomar II and the close shave that the discipline experienced at the hands of the legislative branch of the US government. To his enormous relief, by design, Berg played something of a back seat role in the post-Asilomar events, allowing himself the luxury of returning to his laboratory and once again immersing himself in his true passion — science. However, as presently related, he did not get off scot-free!

Notes

1. Wade N. (1975) Genetics: Conference Sets Strict Controls to Replace Moratorium. *SCIENCE* **187**: 931–935.
2. *Ibid.*
3. Watson JD. (1977) An Imaginary Monster – The Only Danger We Face is the Specter of Untested Regulations. *Bull Atom Sci*, May 1977, p. 12–13.
4. Morgan J, Whelan WH. eds. (1979) *Recombinant DNA and Genetic Experimentation*, Pergamon Press, p. 187.
5. *Ibid.*
6. Letter from David Baltimore to Paul Berg, November 21, 1975. Reproduced with permission from the Department of Special Collections, Stanford University Libraries.
7. Novick R. (1977) Present. Controls Are Just a Start. The Public Will Not Be Protected Until the NIH Guidelines Are Strengthened, Tightened and Made Universal. *Bull Atom Sci*, May 1977, p. 16–27.
8. Letter from Alain Rambach to Paul Berg, January 17, 1977. Reproduced with permission from the Department of Special Collections, Stanford University Libraries.
9. Letter to Alain Rambach, January 26, 1977. Reproduced with permission from the Department of Special Collections, Stanford University Libraries.
10. Interview with Charles Weiner, MIT Oral. History Project, May 1975.
11. *Ibid.*

12. Letter from Janet Goldstein to Paul Berg, January 14, 1977. Reproduced with permission from the Department of Special Collections, Stanford University Libraries.
13. *Ibid.*
14. Letter from Paul Berg, March 15, 1975, The Maxine Singer Papers. [http://profiles.nlm.nih.gov/ps/retrieve/Collection/CID/DJ]
15. Berg P. (2001) Reflections on Asilomar 2 at Asilomar 3: 25 Years Later. *Perspectives in Biology and Medicine* **44**: 183–185.
16. Berg P. (2008) Meetings that changed the world: Asilomar 1975: DNA modification secured. *Nature* **455**: 290–291.
17. Letter from Philippe Kourilsky to Paul Berg, March 1, 1975. Reproduced with permission from the Department of Special Collections, Stanford University Libraries.
18. Letter from Robert Sinsheimer to Paul Berg, March 3, 1975. Reproduced with permission from the Department of Special Collections, Stanford University Libraries.
19. Letter from David Baltimore to Paul Berg, 28 February 1975. Reproduced with permission from the Department of Special Collections, Stanford University Libraries.

CHAPTER TWENTY TWO

The Aftermath

Many hours of ensuing discussion and debate

The Asilomar meeting left in its wake the gargantuan and grinding task of formalizing, implementing and monitoring the rules and regulations that effectively translated the spirit of the conference to the least onerous regulatory practices — the many hours of ensuing discussion and debate that led to federal guidelines, their subsequent revision and their eventual dissolution. Readers interested in blow-by-blow accounts of these milestones are advised to consult the reference section, with particular attention to: *Playing God — Genetic Engineering and the Manipulation of Life*, by June Goodfield, published in 1977; *Biomedical Politics — A Report of the Division of Health Sciences Policy Committee to Study Biomedical Decision Making*, K. E. Hanna (ed.), published in 1991; *Molecular Politics — Developing American and British Regulatory Policy for Genetic Engineering, 1972–1982*, by Susan Wright, published in 1994; *The Recombinant DNA Controversy — A Memoir — Science, Politics, and the Public Interest*, by former NIH director Donald Frederickson, published in 2001 (particularly recommended to gain a sense of the astute role that Frederickson played in these proceedings); and *The Recombinant DNA Debate*, a collection of articles solicited and edited by David Jackson (the former postdoctoral fellow with Berg), and Stephen Stich from the Department of Philosophy at the University of Maryland.

These post-Asilomar efforts were not without their own drama. This was the mid-1970s and government "evils" (notably exemplified by the Vietnam War) were still on the minds of many. As already noted with

Demonstration at a National Academy of Sciences meeting, March 1977.
("The DNA Story" by James Watson and John Tooze, W. H. Freeman, 1981, p. 134.)

regard to the infamous press conference called by Jonathan Beckwith and
James Shapiro in 1969, there was no shortage of public protest in quarters
known for political liberalism, notably the city of Cambridge,
Massachusetts, home of Harvard University — and other freethinking
institutions. The social spillover from media coverage of the Asilomar
meeting is also extensively documented. The Beckwith/Shapiro press conference aside, the city of Cambridge endured a particularly memorable
dose of exposure, lending flamboyant mayor Alfred Vellucci a veritable
circus ring, replete with formal hearings by the Cambridge City Council.

All this prompted Berg to address a lengthy letter to Vellucci, which
ended with a gentle plea for "sanity":

> Would it not make more sense for the Cambridge City Council to
> join with its responsible scientific community in efforts to monitor
> compliance with the guidelines and ensure the safety of the scientists
> and the public at large? Such an action could lead to a partnership for
> progress rather than a conspiracy of repression.[1]

Soon after Vellucci's Cambridge hearings, a second media event exploded.
Liebe Cavalieri, a molecular biologist then at the Sloan Kettering

Memorial Institute in New York, published a highly provocative article in the August 22, 1976 edition of the *New York Times Magazine*. The piece was entitled "New Strains of Life and Death."[2] A full decade later, Norton Zinder (whom the reader will recall served on the Recombinant DNA Organizing Committee), wrote a piece dubbed "A Personal View of the Media's Role in the Recombinant DNA War," for a multi-authored volume entitled *The Gene Splicing Wars — Reflections on the Recombinant DNA Controversy*.[3] He stated:

> Recombinant DNA research was touted on the magazine's cover as a potential holocaust, and inside was an article which set forth many pages of horror. ···.
> There were reverberations all over America. A month later, the article was introduced into the *Congressional Record* by Senator Jacob Javits. Although this garbage was in the media it was not of the media.[4]

Ultimately, the weary Berg simply ignored these sorts of eruptions. He concedes that in the final analysis the moratorium and the regulatory events that followed impeded the progress of research in his laboratory for a period. But he also acknowledges that in the long run this temporary setback benefitted all biomedical research. "For one thing we largely avoided the sort of public opposition that could have easily stopped everything," Berg stated. "For another, I don't know how much work on cloning human genes could have been done right off the bat. We didn't know how to walk, let alone run. We could barely crawl!"[5] Predictably, and to Berg's pleasure, scientists adapted quickly to the imposed do's and don'ts and learned how to deploy recombinant DNA technology safely.

The so-called Berg letter of July 1974 specifically appealed for action by the NIH, prompting the emergence of the NIH Recombinant DNA Program Advisory Committee (RAC). On April 29, 1975, a few months after Asilomar II, Berg forwarded a comprehensive and lengthy report from the Asilomar Organizing Committee to National Academy president Phillip Handler, who after some grumbling agreed to publish a Summary Statement in the June issue of the *PNAS* that year. "The state of the art was to remain relatively stationary through the remaining months of 1975, during which the RAC (a large committee composed of a cross section of participants at the conference) labored to convert the

Donald S. Frederickson.
(http://en.wikipedia.org/wiki/Donald_S._Fredrickson)

Asilomar statements into formal guidelines," NIH Director Don Frederickson stated in his memoir.[6]

On June 23, 1976, 16 months after Asilomar, Fredrickson and his committee completed this gargantuan task and issued final NIH guidelines, which ended up being considerably stricter than those that emerged from the Asilomar II report. DeWitt Stetten, a senior advisor to Fredrickson, defended this course of action as "in no way opening the floodgates, rather a closing of leaks."[7] And Frederickson himself, who by all accounts rendered yeoman service to the cause, proffered that "a core of the history of the NIH guidelines is the story of their revision to modify the required experimental conditions and procedures for the Secretary of the Department of Health, Education and Welfare (HEW), who must agree to the changes."[7]

Even at that late stage of the recombinant DNA story, some (including those from the highest levels of the academic sector) continued to voice opposition to the new technology. In early 1976, Robert (Bob) Sinsheimer, then chairman of the Division of Biology at the California Institute of Technology (Caltech) and a highly respected molecular biologist, wrote a surprisingly naive article pretentiously entitled "On Coupling Inquiry and Wisdom":

> The invention of chimeric organisms bearing recombinant DNA molecules is more than a brilliant scientific achievement; it is the beginning

of synthetic biology. As such it marks a major change in the history of life on this planet. Man-made organisms not derived by the usual evolutionary processes will be introduced into our precious and intricately woven biosphere. Can anyone honestly presume to predict the ultimate consequences? ·· ········
····························· I think we have to believe these creatures will in time enter our biosphere.[8]

This basic message (one that could obviously never be categorically disproved) was delivered on multiple occasions, much to the chagrin and frank astonishment of the biological sciences community, given Sinsheimer's stellar scientific reputation. Not surprisingly, he was the object of considerable scorn and admonition from many of his colleagues for his extraordinary evolutionary hypothesis!

Robert (Bob) Sinsheimer.
(http://www1.ucsc.edu/currents/01-02/art/sinsheimer_robert.)

When Berg received copies of testimony that Sinsheimer presented to the Scientists' Institute for Public Information, he was annoyed and disappointed by many of the statements therein, including those that touched on Sinsheimer's ethics — and chided him in no uncertain terms:

···················· It is neither slander nor innuendo to raise questions

about the depth of your concern when one hears of the following incon-
sistencies! At the same time that you are testifying before scientific, lay
and governmental groups throughout the country that P3-type work* is
too dangerous to be permitted, you have been reported to have taken an
active role in seeking and obtaining funds for the construction of P3 facilities in the Caltech Biology Division. How can you reconcile your recommendation to the *San Diego City Council's Recombinant DNA Study
Panel* not to permit construction of P3 facilities at UCSD (University
of California at San Diego), while you concur that such facilities can be
built at Caltech?[9]

Berg charged Sinsheimer with untiring efforts to persuade other institutions, city councils and the government to reconsider implementation
of the guidelines, while permitting a conspicuous silence in Pasadena and
at Caltech, and even of conducting recombinant DNA experiments in his
own laboratory which, though benign, were the very kind being railed
against by his converts and allies.

Erwin Chargaff, the irascible Columbia University authority on
nucleic acid biochemistry (most famous for documenting the seminal
observation of the corresponding ratio of A and T and G and C in DNA),
was branded as another member of the "lunatic fringe." He is perhaps
equally notorious for his scornful condemnation of Jim Watson and
Francis Crick when introduced to them in Cambridge UK in the early
1950s because of their inability to recall the precise chemical structure of
these bases when asked to do so!

In 1975, Chargaff penned the following purple prose for a piece in a
magazine called *Sciences* (not to be confused with *SCIENCE*):

Knowing that the desire to improve mankind has led to some of the
most horrible atrocities recorded by history, it was with a feeling of deep
melancholy that I read about the peculiar conference that took place
recently in the neighborhood of Palo Alto [several hundred miles from

---

*The recommendations of the Recombinant Advisory Committee included the require-
ment for using barrier facilities of increasing grades of stringency called "P" barriers
(presumably for "prevention"), P4 being the most exacting.

Asilomar!]. At this Council of Asilomar there congregated the molecular bishops and church fathers from all over the world, in order to condemn the heresies of which they themselves had been the first and the principal perpetrators. This was probably the first time in history that the incendiaries formed their own fire brigade. The edict published in due course, which lists various forbidden items, reads like a combined curriculum vita of the conveners of the conference.[10]

This composition was shortly followed by another piece from Chargaff's acid pen in the more recognized journal *SCIENCE*, which included the dramatic portent:

> You can stop splitting the atom; you can stop visiting the moon; you can stop using aerosols; you may even decide not to kill entire populations by the use of a few bombs. But you cannot recall a new form of life.[11]

A similar missive with the identical title, from Francine Robinson Simring of the *Committee for Genetics, Friends of the Earth*, a private non-profit organization, accompanied Chargaff's letter.[12] Both letters predictably elicited a quick response from Berg and Singer. Watson on the other hand refused to retreat an inch from his firm opposition to curtailing

The irascible Erwin Chargaf.
(https://www.google.com/search?q=erwin+chargaff)

recombinant DNA research. As late as 1977, he was still critical of the manner in which the saga was being addressed:

> To say the least, we are in a rotten mess. We face the prospect that our miserable guidelines will be enacted into formal laws. And, their very existence creates fears that DNA research carries a profound threat to human existence. It does not help that some of our "best and brightest" say to each other that they know that Asilomar was a tactical mistake, but then pop-off to Washington to flatter legislators into thinking they are acting wisely in devoting massive attention to DNA.[13]

Notes

1. Letter from Paul Berg to Alfred Vellucci, July 2, 1976. Reproduced with permission from the Department of Special Collections, Stanford University Libraries.
2.   Cavalieri L. *New Strains of Life and Death* -----
3. Zinder N. A Personal View of the Media's Role in the Recombinant DNA War, for a muilti-authored volume entitled *The Gene Splicing Wars — Reflections on the Recombinant DNA Controversy* .
4. *Ibid.*
5. Paul Berg interview by Sally Hughes, Regional Oral History Office of the Bancroft Library at the University of California, Berkeley, 2000.
6. Fredrickson DS. (2001) *The Recombinant DNA Controversy: A Memoir: Science, Politics and the Public Interest.* ASM Press, p. 35.
7. *Ibid.*, p .36
8. Sinsheimer R. (1976) On Coupling Enquiry and Wisdom. *Fed Proc* **35**: 2540–2542.
9. Letter from Paul Berg to Robert Sinsheimer, December 30, 1976, Maxine Singer papers, Library of Congress.
10. Chargaff E. (1976) On the Dangers of Genetic Meddling. *SCIENCE* **192**: 938–940.
11. Chargaff E. (1975) Profitable Wonders, A Few Thoughts on Genetic Meddling. *Sciences*, Aug–Sept, p. 21.
12. Chargaff E. (1976) On the Dangers of Genetic Meddling. *SCIENCE* **192**: 938–940.
13. Watson JD. (1977) In Defense of DNA. *New Republic*, June 5, pp. 11–14.

CHAPTER TWENTY THREE

Legislative and Revisionist
Challenges to Recombinant DNA

You must think me a bloody fool!

Regardless of his emphatic reluctance to remain administratively involved in the fallout of Asilomar, it was difficult (in some instances impossible) for Berg to completely disengage from the extensive regulatory and legislative issues that surfaced in the wake of the conference.

The American government revealed its interest in recombinant DNA early. Just seven months after Asilomar II, Senators Edward (Ted) Kennedy and Jacob Javits, members of the US Senate Committee on Labor and Public Welfare, wrote to Willard Gaylin, president of the Hastings Center at the Institute of Society, Ethics and Life Sciences:

> The recent hearings before the Subcommittee on Health of the Committee on Labor and Public Welfare suggest a growing strain in the relationship between biomedical research scientists and the general public. As legislators who have long been concerned with national health problems, and particularly the support of basic biomedical research, we find disquieting both the tone and direction of some of the current discussion about the public role in the establishment of science policy.[1]

In early 1976, Gaylin shared the Kennedy/Javits letter with a number of key scientists, including Berg, and invited them to participate in a conference to address the US Senators' concerns. Berg was in attendance, but the weary Stanford biochemist emphatically begged off when in the same year he was asked to join the *Committee on Science and Public Policy of the National Academy of Sciences* (CFOSPUP) — a body that convened for four two-day meetings a year for a three-year term.

Berg was, however, persuaded to participate in a National Academy Forum on Recombinant DNA in Washington, DC in March 1977. If nothing else, the event was memorable for the presence of public protesters, who mounted provocative banners in front of the seated academy panel. "We will create the perfect race," one of these banners read, facetiously co-opting an utterance attributed to Adolf Hitler.

Berg, indeed the entire scientific community, desperately hoped to avoid legislation that might hamper deployment of the recombinant DNA technology. But Congressional interest in gene cloning gained considerable momentum following a letter from Javits and Kennedy to President Gerald Ford on July 19, 1976.

> Dear Mr. President:
>
> For several years, the biomedical research community has been engaged in an extremely important debate over the safety of certain types of genetic research. The research involves combining genetic material from different organisms. The technology that permits this type of genetic experimentation, called recombinant DNA research, is revolutionary, and holds the promise of enormous benefits ···.
>
> However, recombinant DNA research also entails unknown but potentially enormous risks due to the possibility that microorganisms with transplanted genes might prove hazardous to human and other forms of life — and might escape from the laboratory. ···.
>
> Given the high potential risks of this research, it seems imperative that every possible measure be explored for assuring that the NIH guidelines are adhered to in all sectors of the research community. We urge you to implement these guidelines immediately wherever possible by executive directive and/or rulemaking, and to explore every possible mechanism to assure compliance with the guidelines in all sectors of the research community, including the private sector and the international community. If legislation is required to these ends, we urge you to expedite proposal to Congress.[2]

In April 1977, the US Senate began drawing up a bill to regulate recombinant DNA research that "proposed establishing within the HEW

[Department of Health, Education and Welfare] an 11-member national commission empowered to license and inspect research facilities and to promulgate safety regulations."[3] The bill was placed on the Senate calendar in October of that year. Intellectual and physical fatigue notwithstanding and realizing the potential for unpleasant face-offs between empowered legislators, Berg was driven to immediate action. On October 17 he sent a telegram to Congressmen Paul Rogers and Harley Staggers urging them not to enact proposed congressional legislation:

> In view of the rapid advance of scientific knowledge by RD [recombinant DNA] and the as yet only speculative nature of the hazards, I regard the approach taken in HR7879 as unwarranted and unnecessary. It is my judgment that the proposed legislation is also unwise since, if enacted, HR7857 will surely inhibit rather than foster basic research on important biological and medical problems; consequently, it will delay the inevitable rewards for the public welfare. In my view the application or modification of already existing mechanisms that guard the public against known hazards is a more prudent way of dealing with research and with any remaining anxieties about RD research.[4]

Around this time, the concerns that had awakened interest from the US Congress began to stir unease in some of the principal scientists involved in the controversy. As a consequence, Waclaw Syzbalski (who, as mentioned earlier, was then a scientifically respected but frequently provocative and contentious professor at the University of Wisconsin) suggested to members of the original MIT Committee that it would be opportune to compose a "revised Berg Letter." His idea was that the revised letter would explain that in the years between Asilomar II and mid-1977, opinions about risks associated with recombinant DNA research had essentially evaporated, especially in light of the issuance of guidelines by a responsible government body — the NIH. "The new letter," Jim Watson and John Tooze wrote in *The DNA Story,* "would give the reasons why, in view of new scientific information and the direction the public debate had taken, the original signatories now felt much less concerned about the conjectural hazards of recombinant DNA experiments than about the real hazards that the legislation pending in Congress might pose as a grave and unnecessary threat to intellectual freedom and scientific progress."[5]

Initially flatly opposed to such a letter, Berg was ultimately persuaded about its potential political utility. On February 6, 1978 he wrote to his colleagues:

Waclaw Szybalski. (http://www.lwow.home.pl/szybalski2003.jgp)

I believe now is a particularly opportune time for us to publish the letter stating our views regarding the need for recombinant DNA legislation. There seems to be a lull in the storm now and I think it would be better for us to have our say prior to, rather than in response to, a new flare up in the debate. Legislators and other government officials appear to be indecisive and hesitant about whether and how to proceed with such efforts and here too our definite statement could be pivotal.[6]

The letter was accompanied by the most current draft of the proposed rewrite of the 1974 "Berg letter." It stated:

In the ensuing four years the discussions, evaluations and experiences concerning recombinant DNA research have strikingly altered our assessment of the risks. More than 250 scientific investigations involving the construction and propagation of many different recombinant DNA molecules have been carried out in the United States without and abroad, with no indication of harm to humans or the environment.[7]

David Baltimore, another signatory of the 1974 Berg letter and a member of both the MIT Committee and the Asilomar Organizing Committee, also supported retrospective reevaluation of that epistle.

We may have so dramatized the issue that we made it possible for a lot of things to happen that wouldn't have happened. I think it's at least arguable that [the many protests] might not have occurred had it not been [for] Paul [Berg], Jim Watson, and me — and the other people who signed that letter. And that, I think, is a serious problem. I mean, that's not a problem in logic, it's a problem again, in sociology. If a group of graduate students had issued that statement it would have not had the impact and so they couldn't have built on it. As it stood, Jonathan King (a member of a panel of scientists that Mayor Alfred Vellucci and other city officials questioned at the peak of the *Science for the People* movement) was able to say — he has said, over and over again, "I didn't raise the problem. They raised the problem." And, "if such a prestigious group raised the problem, there must be a real problem there to be raised." And it's an argument that you can't counter. First of all, you're not generally there to counter it; but secondly, it has a surface validity at least, which is very persuasive. And that's what worries me. I don't think it's wrong what we did. I don't think we could have avoided doing what we did, given how things have developed; it would have been irresponsible to do anything else.[8]

But "as successive alternative drafts of a revised letter were circulated it became apparent that agreement on a text was harder to achieve than it had been in 1974," Watson and Tooze commented. "The sense of urgency was lost as the imminent threat of legislation diminished — and the attempt finally fizzled in the spring of 1978."[9]

David Baltimore.

(http://www.biography.com/improted/iamages/Biography/Images/Profiles/Images/B/ David Baltimore)

Reluctant to let his cause die, in September 1978 Szybalski contacted Berg informing him of an editorial that he had drafted for publication in the journal *GENE*, of which he was editor-in-chief. The editorial, distributed to all the signatories of the 1974 "Berg letter," bore the moniker: "*Present Assessment of the 1974 Letter on the Potential Biohazards of Recombinant DNA Molecules*" and called for its repudiation by the 11 original signers. "Although all well-informed persons are now convinced that the 'potential hazards' of using *E. coli* K-12, as specified in the moratorium letter are nonexistent from a practical point of view," Szybalski wrote, "only one of the signatories [Jim Watson] has clearly repudiated the 1974 letter in public."

The editorial expressed the hope that "a [new] document [can] be created to update and counterbalance the impact of the 1974 letter and to substitute for the joint letter from the signatories ·················

Berg's response, a timeless example of his intolerance and disdain for blatant self-promotion, was unsparing in its condemnation.

Dear Waclaw,

You must think me a bloody fool! So let me be quite frank. My present views on the risks of recombinant DNA research are on the public record and quite clear. Even if I was looking for a way to make them clearer I would ignore your arrogant and crude "invitation." I regard your scheme and proposed actions as clumsy and irresponsible, not so much for what you are trying to achieve, but for the heavy-handed and insulting way you've gone about it. I wouldn't be party to it if my reputation depended on it.

GENE purports to be a scientific journal. It is struggling to make a reputation and become an attractive focus for the exciting science that is yet to come. Now, by your actions, distortions, outright misstatements of fact [and] slick innuendos bordering on plain old-fashioned yellow journalism, will become part of its content. ·································[10]

About eight months later, the chief editor of a little known (at least to this author) German journal called *UNSCHAU in Wissenschaft und Tecknik* stuck his head in the proverbial oven by resurrecting withdrawal of the original "Berg letter" in which he *instructed* [author's italics] Berg to respond to a series of inane questions. Most recipients of this sort of letter would likely mutter an appropriate profanity and throw it in the trash. A month later, Berg responded with the following opening sentence: "You will excuse me, I hope, if I say that I am astonished at your ignorance of what has transpired in the last five years of the recombinant DNA issue."[11]

His best intentions notwithstanding, Berg was called to testify before the Senate Subcommittee on Science, Technology and Space, chaired by Senator Adlai Stevenson IV. In his prepared testimony Berg addressed "the opportunities and important biological problems that recombinant DNA research can help solve."[12] He then turned to the risks:

> Enacting legislation to govern the content and methods of scientific en-quiry would be unprecedented and probably unworkable. In my view legislation of the type that has so far been proposed would inhibit basic research on important biological and medical problems. The rules, pro-cedures, and penalties are predicated on assumptions that will surely change, thereby making it difficult and cumbersome to adjust to the changing information, ideas and opportunities. I believe that legislation could stultify the creativity and initiative that has characterized the de-velopment of the recombinant DNA technique; it could also discourage and disillusion young scientists from entering this field.[13]

Not wanting to be excluded from the rising tide of politically motivated concern, several state legislative bodies also raised their voices about recombinant DNA. In 1977, Berg was informed by Barry Keene, chair-man of the California Assembly Committee on Health, that he was spon-soring legislative bill #AB757, the *California Biological Research Safety Act.* Perhaps Berg got up on the wrong side of the bed that morning. More likely, he was worn thin by the self-serving manner in which scientifically naive politicians were jumping into the fray. His response to Keene repre-sents another fine example of Paul Berg on the warpath! He wrote:

> To eliminate any confusion about where I stand let me speak quite frankly and bluntly. I do *not*, as you seem to have implied to one re-porter at your press conference, support AB757. On the contrary I am unalterably opposed to its intent, astonished by its simplistic and naive assumptions, insulted and dismayed by its provisions and committed to oppose its enactment. ⋯⋯⋯⋯⋯⋯⋯

One of the things that pains me greatly is that I accepted your assurances that there would be an open-minded, careful analysis of the problem, followed by wide-ranging consultations about how to deal with the issues. Instead, the matter has become politicized, charged with hysteria and riddled with intrigue, distrust and cynicism, all of which makes a resolution of the questions that face us more difficult. I had hoped that I could work *with* you rather than *against* you [author's italics]. But if you persist in pressing this bill it will certainly be the latter.[14]

A silver lining to the gathering legislative clouds in Washington emerged in March 1978, when Senator Ted Kennedy introduced an amendment to the Senate bill that signaled an altered perception of risks associated with recombinant DNA technology. As Kennedy no doubt intended, the bill died in the Senate — and with it any further regulatory considerations by the US government. A huge sigh of relief emerged from all concerned about legislative interference.

By this time, Berg had lost much of his messianic fervor, especially when monetary gain inevitably raised its ugly head. "I must have testified several times," he stated. "But my cynicism arose when the entire investigation disappeared as soon as Genentech went public (in 1980). I assume that was when people recognized the enormous commercial potential of recombinant DNA technology."[15]

On the heels of Senator Kennedy's withdrawal from a potential ruckus, a general sentiment began to grow to revise and relax the federal guidelines. Between early September and early November of 1978, Berg had extensive correspondence with Don Fredrickson. In one of his missives he composed a lengthy congratulatory letter to the NIH director for his efforts:

.... [H]ow far we have moved since the issue of potential risks associated with recombinant DNA research was first raised publicly. In spite of the occasional ferocity, stridency and all too frequent politicization of the discussions, features that turned me and others off to continuing involvement, you and your colleagues have persevered, steered a reasonable course while peppered with foolish and intemperate advice and arrived at proposals, which in the main are workable and a basis for further evaluation.[16]

Surprisingly, as late as May 1980, David Baltimore, who had previously voiced unwavering opposition to government oversight, challenged Berg about the astuteness of phasing out the painfully and laboriously negotiated recombinant DNA guidelines.

> I am curious about whether you now feel that there is no conceivable way that any type of recombinant DNA experiment could be hazardous [a view that Berg had recently expressed in his impatient correspondence with Waclaw Szybalski]. I am personally not satisfied that, for instance, use of a wild-type *E. coli* strain as a recombinant DNA host is advisable. Similarly, there are a wide range of organisms that, because of their ecological distribution, would seem to be best not tampered with. ....
>
> If you are convinced that no hazard is conceivable then I completely agree that the Guidelines should be downgraded to a voluntary advisory position because then one would only be taking reasonable precautions, not worrying about any defined problem. If, however, you think that problems might yet be out there, then I think that RAC [Recombinant DNA Advisory Committee] is serving its function as it stands and has, to a large extent, phased itself out of non-controversial areas of recombinant DNA work and left itself a part of the more complicated issues.[17]

In his response to Baltimore, Berg wrote:

> First, let me say that I am not totally persuaded "that there is no conceivable way that any recombinant DNA (RD) experiment could be hazardous." But I find it increasingly difficult to accept the maintenance of a non-trivial government bureaucracy to contend with that mythical possibility. After all, we do not fling the same challenge toward virus research or other "potentially hazardous" laboratory activities. In those instances, society and the government have accepted individual and institutional assurances for safe practices and culpability for negligence or deliberate violations. Shall we forever be bound by the precedent of RAC [Recombinant Advisory Committee] and be obliged to that format because we can't say that there in *no* conceivable way for RD experiments to be dangerous?[18]

A year later Berg had no reservations in sharing these views with Bill Gartland, head of the Office of Recombinant DNA at the NIH, in support of a statement from Baltimore and Allan Campbell, which appeared in the Federal Register in late March, 1981. "Lest there be any doubt about my views," Berg wrote to Gartland, "let me be quite direct. I strongly favor their suggestion and have done so for some time."[19] In December 1980, the relaxed revisions to the guidelines were made official.

With the passage of time, Berg occasionally felt the imperative to justify these suggested changes in public. In response to a question about the dangers of recombinant DNA experimentation following a presentation he made to a public forum in the Bay Area in June 1981, he stated:

> Those of us expressing concern with the early research did not say that there was a danger, but that we didn't *know* [author's italics] if there was a danger. We tried to appeal to the scientific community to slow down so that we may have a period to examine the question. We spent about five years analyzing that question and the current consensus is that there is little or no risk in this type of research.[20]

Curiously, the surge of social conscience in the wake of Asilomar received little (if any) analysis by professional sociologists. No authority on human behavior seems to have addressed the question: Why were (some) scientists, a group historically infamous for their apathy concerning possible social fallout of their discoveries, suddenly apprehensive about genetic engineering? Perhaps the issue didn't merit much introspection. After all, in the final analysis, many observers cynically concluded that the scientists were largely motivated by anxiety that government intervention would hinder progress in their laboratories. But in keeping with the theme of sociological implications of the controversy, Berg received an enthusiastic 6500-word term paper by a thoughtful freshman undergraduate at Duke University.

> In two drastically different times, with drastically different players, the Manhattan Project proceeded but the recombinant DNA technique was buried under a moratorium. The fates of the atomic bomb and rDNA

intertwined with the conviction of Vannevar Bush and Paul Berg respectively. ··· The postwar controversy surrounding the atomic bombs coupled with demands for social responsibility arising from domestic dissatisfaction and unrest shaped the mentality in which Paul Berg arrived at the moratorium. The atomic bomb and rDNA developed along contrasting trajectories because of the differences of political climates and their effects on Bush and Berg and those from whom they sought advice.

More recently, Berg lent his biological expertise to a group led by George Schultz, Secretary of State under President Ronald Reagan for almost seven years. A former advisor to several presidents of the United States, and presently a Fellow at the Hoover Institute at Stanford University, Schultz leads a group that aims to develop policy recommendations to address climate/energy issues facing the nation. The taskforce includes economists, engineers and others involved in energy issues. Schultz invited Berg to join so that he (Berg) could inform the group on the likelihood that biofuels might provide an alternative to petroleum-based gasoline. "That gave me the incentive to learn more about synthetic biology or synthetic genomics which, after all, might be viewed as recombinant DNA on steroids," Berg stated.[21]

Notes

1. Letter from Senators Edward Kennedy and Jacob Javits to Willard Gaylin, October 29, 1975. Reproduced with permission from the Department of Special Collections, Stanford University Libraries.
2. Letter from Senators Jacob Javits and Edward Kennedy to President Gerald Ford, July 19, 1976. Reproduced with permission from the Department of Special Collections, Stanford University Libraries
3. Genetic Engineering, Human Genetics, and Cell Biology: Evolution of the Technological Issues By Congressional Research Service, p. 38, 2002.
4. Telegram for Paul Berg to Paul Rogers and Harley Staggers dated October 17, 1977; Paul Berg, personal communication, October, 2012.
5. Watson JD, Tooze J. (1981) *The DNA Story: A Documentary History of Gene Cloning.* W. H. Freeman and Company, p. 251.

6. Letter from Paul Berg to various colleagues, February 6, 1978. Reproduced with permission of the Department of Special Collections, Stanford University Libraries.

7. Document accompanying Berg's February 6, 1978 letter. Reproduced with permission of the Department of Special Collections, Stanford University Libraries.

8. Letter from David Baltimore to Paul Berg. ------- Reproduced with permission from the Department of Special Collections, Stanford University Libraries.

9. Watson JD, Tooze J. (1981) *The DNA Story: A Documentary History of Gene Cloning.* W. H. Freeman and Company, p. 251.

10. Letter from Paul Berg to Waclaw Szybalski, October 10, 1978. Reproduced with permission from the Department of Special Collections, Stanford University Libraries; Paul Berg, personal communication, October, 2012.

11. Letter to H. Schultze, June 7, 1979. Reproduced with permission from the Department of Special Collections, Stanford University Libraries; Paul Berg, personal communication, October 2012.

12. Testimony by Paul Berg to US Congress, November 2, 1977.

13. *Ibid.*

14. Letter from Paul Berg to Barry Keane, March 28, 1977. Reproduced with permission from the Department of Special Collections, Stanford University Libraries.

15. ECF interview with Paul Berg, February 2012.

16. Letter from PB to Donald Frederickson, September 20, 1978. Reproduced with permission from the Department of Special Collections, Stanford University Libraries.

17. Letter from David Baltimore to Paul Berg, May 28, 1980. Reproduced with permission of the Stanford Department of Special Collections, Stanford University Libraries.

18. Letter from Paul Berg to David Baltimore, July 2, 1980. Reproduced with permission from the Department of Special Collections, Stanford University Libraries.

19. Letter from Paul Berg to Bill Gartland, April 22, 1981. Reproduced with permission from the Department of Special Collections, Stanford University Libraries.

20. *The Commonwealth*, June 22, 1981, p. 173.

21. ECF interview with Paul Berg, February 2012.

CHAPTER TWENTY FOUR

Asilomar II — Lessons Learned

The "social contract" between science and the public has served us well for a long time

The Asilomar paroxysm and the immediate events that followed invited other retrospective analyses. In 1981, Joann Ellison Rodgers, an award-winning science journalist, magazine writer, book author and editor provided a perceptive analysis of the Asilomar Conference in an article entitled "Asilomar Revisited," in which she addressed issues not previously identified:

> To begin with, short of biological weaponry, Asilomar was probably the first popular American mass experience with biological risk identification and potential government–public regulation of biology; virtually all others, from X rays to chemical waste disposal, have had to do with so-called hard technology. As such, it is a landmark in the history of science, an experience which has forever altered public perception of science and science policy in and out of biology. While in many respects Asilomar was a logical outgrowth of standard operating procedures for scientists, Asilomar also entailed an unfamiliar sense of urgency — speeding up and formalizing a process that normally operates slowly and informally. ⋯⋯⋯⋯⋯⋯⋯⋯
> ⋯⋯⋯⋯⋯⋯⋯⋯⋯⋯⋯⋯⋯⋯⋯⋯⋯⋯⋯⋯⋯⋯⋯⋯⋯⋯⋯⋯⋯⋯⋯⋯⋯
>
> [Charles] Weiner (oral historian and director of the MIT Recombinant DNA History Project) admits that there are problems associated with interpreting historical events too soon. "But with this issue, events are moving so fast that it might be too late."

It's now clear from the reflections of key participants and a review of the archive that Asilomar overwhelmed its organizers because of the peculiarly American social, scientific and political environment in which it occurred. Money and talent in prodigious amounts had put American scientists in the forefront of molecular biology by the late 1960s. In the process, however, molecular biologists had reduced life to *tour-de-force* exercises in physics and chemistry; they had little to do with organisms, pathogenic or otherwise.[1]

In 2010, the Asilomar model was also co-opted by a group interested in debating climate control. When this assembly organized the Asilomar International Conference on Climate Intervention Technologies, it approached Berg for help and advice. He willingly addressed two key administrative entities; the Climate Response Fund and the Climate Institute on conference structure, processes for guideline development and post-conference interaction with the scientific community. Berg even went so far as to attend the conference on climate intervention technologies and to serve as honorary chair.[2]

More recently, he composed an unpublished essay that formulated comparisons between the level and extent of public concern about recombinant DNA and the more recent apprehension (or lack of it) regarding the highly dangerous bird flu virus H5N1. H5N1 virus is highly transmissible in some animals (notably birds), but fortunately, given its extreme virulence, not in humans. However, two research groups independently demonstrated that deliberately mutating the virus in the laboratory can in fact allow its transmissibility in humans.

Controversy began in late 2011 after the US National Science Advisory Board for Biosecurity (NSABB) requested that one of the research teams significantly modify a scientific article they intended to submit for publication through the usual peer review process. The board did so out of concern that bioterrorist groups could potentially misuse the published research for harmful purposes. Some experts concerned about censorship of scientific publications advanced the contrary opinion that the data gained from these studies might be particularly helpful in reducing the potential for spontaneous mutations that could conceivably trigger transmissibility in humans — and a resulting pandemic. In his essay

Berg emphatically stressed some of the lessons the Asilomar experience taught him — lessons that beg the crucial imperative for scientists to continue to sit down together and engage in dialogue whenever situations arise that may threaten public safety, and to inform the public at the earliest reasonable time.

Berg addressed the H5N1 virus issue in a characteristically thoughtful manner. "The most important lesson I learned from the recombinant DNA debate and its outcome is how important it is to establish trust with the public about what one knows, what one doesn't know, and what needs to be learned," he stated. "An edgy public is something to be avoided because in any society there will be some who will try to exploit the concerns and stymie efforts to move on. In that regard it's appropriate to ask if the manner in which the recombinant DNA issues were dealt with are relevant for dealing with the concerns raised by the creation of the more transmissible H5N1 virus?"[3] He wrote:

> The concerns surrounding the possible dangers of permitting the publication of a scientific paper on recent research on H5N1 virus are a good case in point: Specifically, if the public was made fully aware of why publishing the results of research of the modified flu virus was considered a risk [of bioterrorism] they might have better understood why discussions on how to proceed had to be conducted behind closed doors. In this regard, it is also relevant to ask whether the public realized that the initial decision not to publish curtailed the ability of the scientific population to sort out the genetic determinants of the enhanced transmissibility of the modified H5N1 virus, and possibly also delineating regions of the viral genome that confer pathogenicity. I was frankly surprised by how subdued the public's concerns were about accidental release of the modified virus and the possible dire consequences thereof.
>
> With respect to both the recombinant DNA saga and the more current H5N1 concerns, the inadvertent release of a potentially dangerous agent has to be of equal or even greater concern. We should be vigilant to the possibility that in both instances inattention to security or accidental release could lead to a more serious outcome. History

and the laws of probability are replete with examples of accidental release of dangerous agents. If work in this general arena is to continue because of the overriding interests of potential gains to be realized for human health, the public needs to be assured that leaks by any means are as preventable as technology can provide.

In the recombinant DNA experience and now in the modified H5N1 flu virus controversy, I have concluded that hubris, the very antithesis of prudence, runs high among scientists. There is an incredible ability to rationalize or even ignore potential risks in their work. But being able to predict the outcome of experiments with 100% certainty is beyond any of us. Indeed, it is the unexpected result we are always alert for. However, in pursuing our research, we owe it to humankind to do no harm.

Berg concluded his essay on a cautionary note:

We are pretty much agreed that the "social contract" between science and the public has served us well for a long time. But those shared values are undergoing new strains in dealing with the conduct and consequences of biomedical research. Otherwise legitimate investigations aimed at the acquisition of new and fundamental knowledge, and possibly the development of new life-saving therapies, are being increasingly scrutinized and occasionally resisted by the polity. Confounding that scrutiny is that it is too often energized by religious, ideological and cultural concerns, concerns that interpret "risks" in a very different light.[4]

Notes

1.   Rogers J. (1981) *Asilomar Revisited. Mosaic*, Jan/Feb 1981, 19–25.
2.   http://climateresponsefund.org
3.   ECF personal communication from Paul Berg, October 2012.
4.   *Ibid.*

PART V

CHAPTER TWENTY FIVE

The Nobel Prize in Chemistry

Let's be absolutely frank about this

In 1980, the Nobel Foundation acknowledged contemporary DNA technology in its award in chemistry. As mentioned earlier, Fred Sanger at the MRC Laboratory of Molecular Biology in Cambridge, England, and Walter (Wally) Gilbert at the Cambridge on the other side of the Atlantic shared half the monetary award for highlighting DNA sequencing technology. The other half was awarded to Berg.

When formally announcing the award to Berg in a press release, the Nobel Foundation declared:

> The Royal Swedish Academy of Sciences has decided to award the 1980 Nobel Prize in chemistry by one half to Professor Paul Berg, Stanford University, USA, for his fundamental studies of the biochemistry of nucleic acids, with particular regard to recombinant-DNA.[1]

Though not explicitly stated, this sentence clearly specifies that Berg was being recognized for contributions to the discovery and role of tRNAs in protein synthesis — and for contributions to the development of recombinant DNA technology. A subsequent expanded announcement focused exclusively on his recombinant DNA work:

> Berg was the first investigator to construct a recombinant-DNA molecule, i.e., a molecule containing parts of DNA from different species, e.g., a chromosome from a virus combined with genes from

a bacterial chromosome. His pioneering experiment has resulted in the development of a new technology, often called genetic engineering or gene manipulation.[2]

"Conceivably, my involvement in calling attention to the potential risks and in organizing Asilomar II, which led to guidelines for the use of recombinant DNA technology may have contributed to my selection by the Nobel committee," Berg commented. "I believe the choice of recipients acknowledged the advance in DNA sequencing and the initiation of the recombinant DNA technology as complementary developments, each in their own way contributing to a transformational advance in genetics."[3]

If they reside in the United States, Nobel laureates are typically informed of their newly won status with an early morning phone call — sometimes *very* early in the morning! One assumes that these calls are timed to eclipse early morning media releases. Berg learned of his 1980 award when the shrill ringing of his home telephone awakened him on the morning of October 10, 1980*.

Fatefully, Berg's father was then critically ill, so the thought uppermost in his, Millie's, and his son John's minds concerned the state of Berg's father's health. Though greatly relieved to hear a voice other than that of a family member communicating bad news, Berg was initially unable to recognize the voice on the other end of the phone, a voice that excitedly blurted: "Have you heard the news?" "What news?" Berg asked, now recognizing the caller as Arthur Kornberg. "You have just won the Nobel Prize in Chemistry," Kornberg emphatically replied.

It is presumably not unheard of for a Nobel awardee to learn of the joyful news from a sleepless friend or colleague. In this case the Nobel Committee on Chemistry was in fact unable to contact Berg directly because his home phone number was unlisted. Providentially, Arthur Kornberg's son, Tom, who happened to be in his office at the crack of dawn, heard the announcement on his radio and immediately

---

* All except the Peace Prize are announced around October 10, the date of Alfred Nobel's birth.

communicated the news to his father. In the midst of frantic scurrying in the Berg household to dress and greet a swarm of news reporters crowding the front porch, Berg alarmingly recalled that that very day he and Millie were scheduled to leave for Toronto, Canada, for festivities in honor of Berg's recent receipt of the Gairdner International Award, yet another of the many prizes and awards heaped on him.

He was in a quandary. He knew full well that Stanford University would soon be scurrying to organize meetings of recognition and congratulations that day — which he could scarcely avoid. What to do? He called Louis Siminovitch, his host in Toronto, to inform him that he was unable to attend the Gairdner festivities that day, an announcement that predictably prompted Siminovitch's unfettered surprise — and consternation!

Finding it embarrassing to inform Siminovitch about just having heard about the Nobel prize, Berg mumbled that something unexpected had come up. Inevitably, this response failed to placate his Canadian colleague (whom he knew well), so following several further feeble attempts to assuage Siminovitch's distress, Berg bit the proverbial bullet and informed him of the startling events of the previous few hours. "I was simply unable to get those words out" Berg commented in oblique reference to his self-consciousness about awards, a characteristic well known by those close to him. Of course when the truth finally emerged, Siminovitch happily reassured Berg that the Gairdner Award celebrations would be postponed for 24 hours.

Thus began weeks of celebration and hectic preparation for the visit to Stockholm a short few months ahead. Among other logistical demands was the need to purchase clothes more suitable for winter in Stockholm and to prepare his Nobel lecture presentation. The weeklong festivities in December also included the presentation of scientific seminars at several Swedish universities.

The ceremonies that embrace the formal award and the banquet that follows are traditional highlights of Nobel week. Though each impending laureate is allowed to designate a number of guests, few lesser mortals have the privilege of personally experiencing Nobel week. So readers might enjoy a quick peek behind the curtains of this exclusive and lavish event, one that historically far outweighs any other scientific award in its distinction.

"Nobel week is essentially consumed by a series of functions, receptions, interviews and dinners," Berg related. "Each laureate and his/her spouse is provided with a driver and a trained personal escort, usually from the Swedish Foreign Service, to be certain that one is properly dressed for whatever special event was next; that one's tie was tied correctly; that one arrived at the right place at the right time — and so on."[4]

"We never used the chauffeur," Berg scoffed. "We don't typically drive around with a chauffeur, so why start now? Additionally, when the family dined out for a few meals alone we simply strolled into the old part of Stockholm — an easy walk from the hotel." John Berg enterprisingly met up with the daughter of one of the physics prizewinners! The pair had a marvelous time being chauffeured to a rock concert and other Stockholm attractions.

The Nobel award ceremony is tightly orchestrated and rehearsed. The laureates are assembled in a private room and are instructed as to how to enter the hall and take their seats — to the strains of stirring music! The stage holds a row of chairs at one end for the laureates; the Royal Family is seated at the other end close to a group of distinguished Swedish scientists. The chairperson of the committee that awards each category of prizes (physics, chemistry, physiology or medicine, economics and literature) reads a citation in Swedish and English. The King of Sweden and

The 1980 Nobel laureates in Chemistry. From left to right, Frederick (Fred) Sanger, Berg and Walter (Wally) Gilbert. (Courtesy of Paul Berg.)

individual awardees come to center stage and the royal personage presents each with a medal and a leather binder containing a distinctive drawing.

Berg had previously met King Carl XVI Gustaf soon after the so-called Berg letter had become public. On that occasion the pair had exchanged comments after Berg's lecture at the Karolinska Institute. "So nice to see you again," the king warmly commented when he handed Berg his Nobel medal. When the award ceremony is completed friends and family stream onto the stage and for a brief period the highly ceremonial air degenerates into a sort of free-for-all.[5]

The award ceremony is followed by a sumptuous formal banquet at the Stockholm city hall, a gorgeous facility that accommodates 1300–1500 people for a sit-down dinner. The laureates enter the room to a blast of trumpets — each with a female escort on one arm. At some point during the dinner a single laureate for each prize category offers a few remarks; *few* being the operative word. Berg was the anointed spokesperson for the three laureates in chemistry. Sanger was invited to make additional

Paul and Millie Berg at the Nobel ceremonies.
(Courtesy of Paul Berg.)

remarks in recognition of the fact that this was his second Nobel prize in chemistry, the first having been awarded in 1958. Everyone then retires to a huge and elegant ballroom situated above the dining hall, for partying and dancing that continue until the wee hours.[6]

Nobel week includes a dinner at the Royal Palace, attended exclusively by the King, the Nobel laureates and their spouses, and members of the diplomatic service. At some point the Swedish King quietly moves to one end of the room from which he silently dispatches an administrative person to escort the laureates — one at a time for a few private words. In December 1980, the recombinant DNA debate was at its height in Sweden, and Berg and King Gustaf enjoyed an extended conversation on the topic. When it was the physicists' turn to meet with the royal personage, the King asked Berg to remain a little longer because "he was uncertain about how he would fare in a discussion about violations of symmetry principles and the existence of matter and antimatter, the subject of the prize in physics!"[7]

After the weeks of preparation for the visit to Stockholm, not to mention the intense celebratory week there, the Berg family was more than ready to return to Palo Alto, California — and the quiet and peace of their home. "A standing joke among Nobel laureates is that when one returns home one is advised to don a pair of jeans and an old T-shirt — and scrub the kitchen floor; a quick way to return to the real world," Berg stated. "But, in reality, life thereafter did change. The self-consciousness I experienced in telling Siminovitch that I had received the prize remains, but I've become accustomed to dealing with the celebrity issue. It is not uncommon to be invited to dinner and discover that the person sitting next to you has been informed: 'You will be sitting next to a Nobel prize winner at dinner.'"[8] One is therefore left with the conclusion that in the main celebrity status is enjoyed more by those who rub shoulders with superstars than by the celebrities themselves!

****

As mentioned earlier, there is little question that Stanley Cohen and Herb Boyer merited recognition for their contributions to advancing the new technology of gene cloning — not to mention the challenging problem of bacterial antibiotic resistance. Indeed, some, especially those not intimately familiar with the unfolding of the recombinant DNA saga, have expressed the view that the two plasmid authorities should have enjoyed a share of the 1980 Nobel Prize in Chemistry — or in a different category, either in 1980 or subsequently. But the workings of the various Nobel committees through which the annual awards wend their convoluted way are delightfully unpredictable. Aside from the challenge of recognizing both a particular field of science and those who made seminal contributions to it, as pointed out earlier, Nobel committees are hamstrung by the rigid rule that no more than three individuals may share in any given award and that awardees must be living. Perhaps the most that can be expected from the challenging charge of selecting laureates from among the many worthy scientists, literary figures, economists and those who make exceptional contributions to promoting peace in a troubled world, is that the Foundation avoid overtly embarrassing choices — as has transpired in a few instances.

In 2009, interviewer Sally Hughes confronted Cohen with a forthright question: "Was there some feeling on your part that it would have been well to have been included?" Cohen was equally forthright with his answer. His response is presented here essentially verbatim:

> I did feel that if a Nobel Prize was given in this area, the discovery that genes could be transplanted to and propagated and cloned in a foreign host was the key discovery; and that discovery came from the work that my lab and Boyer's had done. ⋯⋯⋯⋯⋯⋯⋯⋯⋯⋯⋯⋯ [I also] felt that the joining of DNA ends had been done by others prior to Berg's work, and that once it was known that DNA ends — synthetic or natural — held together by base pairing of nucleotides could be ligated *in vitro*, there was not something special about being able to ligate the ends of DNA molecules that were taken from different species. DNA ends are DNA ends, regardless of the source. But DNAs from different sources have differences in overall sequence and nucleotide composition, and the key question had been whether the sequence of nucleotides in a DNA molecule taken from a different species would allow its propagation in a foreign host. ⋯⋯⋯⋯⋯⋯⋯⋯⋯⋯⋯⋯ I thought that he might share in any Nobel Prize that might be given in this area,

but on the other hand, I also felt that our discovery that novel DNA molecules constructed *in vitro* could be propagated and cloned didn't depend in any way on Berg's earlier work or use any information his lab had produced. If his work had not been done, there would not have been a difference in the development of the field, and it was hard for me to imagine that Berg would receive a prize for an experiment that he did not do. ···

I was surprised a few days later to receive a morning call at home from a reporter asking me what I thought about the fact that Paul Berg had just won the Nobel Prize. "Really?" I said, "The Nobel Prize in what? Honestly, my first thought was that it was possibly the Peace Prize for Paul's role in encouraging scientific self-control in an important area of biomedical research. I asked, "What contributions were cited by the Nobel committee?" And I was told that the prize would be given in chemistry for Paul's "contributions to the biochemistry of nucleic acids with special regard to recombinant DNA." Paul had made a number of earlier important contributions to nucleic acid chemistry, but I didn't understand what was meant by the vague statement: "with special regard to recombinant DNA," and I didn't know what specific achievement was being referred to by that statement. ··················· But I was very, very surprised. Later that morning, to get ahead in the story, is when I learned that the contribution being recognized was not DNA propagation or cloning, but rather simply the biochemical process of joining together DNA ends. But recombinant DNA technology was much more than end-to-end joining of DNA molecules; it was the propagation of DNA in a foreign host, and that was not done by Berg. ····················· ·································

I think that award committees are human like everyone else, and they make decisions based on the information they have. The prize wasn't awarded in medicine by the Karolinska Institute, as usually happens with Nobel Prizes related to biological discoveries, and I have to assume that the Nobel committee that selected Paul for the chemistry prize, a different group of scientists working in the area of chemistry believed that his contribution was the seminal one. ························ ·························

There are people out there who feel that perhaps Herb and I should have received that recognition, but that's now ancient history. ···············

[T]he Nobel Prize issue bothered me for a long while, to be honest about it, but it's something I've gotten over. ···························· And there has been a lot of other recognition that Herb and I have received.[9]

When Berg added the trophy *Scientist of the Year* (awarded by the magazine *Discover* since 1964) to his list of prizes and awards, a feature article entitled "Paul Berg and the Ultimate Technology" proffered:

> In the intensely competitive world of genetic engineering there are certainly researchers who will readily suggest others who were as much entitled to the 1980 Nobel, as was Berg — or, for that matter, who should have been named *Scientist of the Year*. Stanley Cohen, a molecular geneticist at Stanford whose accomplishments in recombinant DNA studies have won acclaim, is a major example. But no one would deny that Berg, having won the Nobel, deserved it.[10]

After all is said and done, it was Janet Mertz and Ron Davis in Berg's laboratory who made the critical discovery that cleavage of DNA by the *Eco*R1 enzyme enables two different DNAs to be joined. Cohen and Boyer discovered that appropriate molecules joined in that way can be propagated, i.e. *cloned*, in bacteria.

The accolades accorded to Berg and Cohen profited Stanford University's reputation mightily. The year 1980 was a banner one for Berg, who pocketed three prestigious prizes — the Nobel prize, the Gairdner Foundation's International Award and the Lasker Foundation Basic Research Award. In the course of the next several years he was elected a Foreign Member of the French Academy of Sciences and of the Royal Society of England. He also received the National Medal of Science and the American Association for the Advancement of Science (AAAS) Scientific Freedom and Responsibility Award — the last mentioned specifically for his contributions to resolving the recombinant DNA debate. As for Cohen and Boyer, in addition to the 1980 Lasker Award that the two shared with Dale Kaiser and Berg, the pair have to date garnered their own impressive covey of prizes — some with substantial monetary awards!

By way of a final word on the debates that regularly ensue following the annual announcement of the Nobel prizes, to this author's knowledge Herb Boyer never publically entered the fray concerning the 1980 Nobel Prize in Chemistry. When, in 2009, interviewer Jane Gitschier asked him: "Many people have speculated about why it is you and Stan Cohen have never won a Nobel Prize, but I don't know that you've ever talked about that publicly. Are you going to address this in your memoir?" "I will, and I don't mind talking about it with you," Boyer responded. "It is not for me to decide whether I should or should not win a Nobel prize. I've received many prizes and honors and I am indeed grateful for the recognition. You can imagine from what I've said about my boyhood [earlier in the interview] that I never would have expected to do what I've done. I wanted to do something important — I didn't know what it would be, but I planned to work hard and see what I could do. I had no foresight that this would be what it was."[11]

Notes

1. http://nobelprize.org/nobel_prizes/chemistry/laureates/1980/press.html
2. *Ibid.*
3. ECF interview with Paul Berg, October 2012.
4. *Ibid.*
5. ECF interview with Paul Berg, August 2012.
6. *Ibid.*
7. *Ibid.*
8. *Ibid.*
9. SC interview by Sally Hughes, Regional Oral History Office, the Bancroft Library, University of California, Berkeley, 2009.
10. Duncan DE. (1981) *Discover Dialogue: Biochemist Paul Berg.* Discover, January 1981, p. 7.
11. HB interview by Jane Gitschier, September 2009. Wonderful Life: An Interview with Herb Boyer. *PLoS Genet*, September 2009, e1000653.

CHAPTER TWENTY SIX

Commercializing the Technology

It may completely change the pharmaceutical industry's approach

A few months prior to his July 18, 1974 piece in the *New York Times* that publically addressed the brewing recombinant DNA technology controversy, science writer Victor McElheny also rendered an article in the *Times* featuring the *PNAS* paper by John Morrow, Stan Cohen, Herb Boyer and others, on their experiments with frog DNA. *The Stanford Daily* (the university's daily student newspaper) also highlighted this news and included a quote from Nobel Laureate Josh Lederberg, who, in referring to the technology as "a major tool in genetic analysis," prophetically suggested: "it may completely change the pharmaceutical industry's approach to making biological elements such as insulin and antibiotics."[1]

Details of the commercialization of recombinant DNA technology and the parallel ascendance of the biotechnology industry are extensively addressed in several books: *The Golden Helix: Inside Biotech Ventures*, by Arthur Kornberg (1995); *Biotech — The Counterculture Origins of an Industry*, by Eric Vettel (2006); *Genentech — The Beginnings of Biotech*, by Sally Smith Hughes (2011); and *The Recombinant University*, by Doogab Yi (personal communication). Events that transpired at Stanford merit brief mention here.

When articles touting the collaborative efforts at Stanford and the University of California at San Francisco appeared in the *San Francisco Chronicle* and *Newsweek*, Bob Byers, director of the Stanford News Office, forwarded them to Niels Reimers, in charge of Stanford's Office of Technology and Licensing.

In the very early days of protecting discoveries, a number of American universities, including Stanford, typically forwarded technology disclosures to a New York-based entity called the Research Corporation. "After a few months they'll respond, and then you take it from there," was the rejoinder that Niels Reimers received when he enquired about the handling of technology disclosures shortly after his arrival at Stanford in 1968. A quick look at past efforts revealed that the university had earned the princely sum of ~$4,500 between 1954 and 1957! Singularly unimpressed, Reimers initiated a formal technology-licensing program. This initiative yielded ~$55,000 in its first (partial) year and $75,000 the following year. By 1997, this figure had risen to ~$40 million.[2]

The early publicity surrounding the emergence of recombinant DNA technology initially prompted little thought among scientists about commercial gain from their work. For one thing, though financial motives are of obvious relevance to the patent holder, few scientists then understood that from a regulatory point of view the patenting process was specifically designed "to spur industry to develop inventions into socially useful products by providing a period of protection — not to generate financial gains."[3] Even among those who appreciated this nuance there were many who (incorrectly) believed that financial benefits that might accrue from their research would enrich the university involved, not the inventor. Then too, in the early 1970s, many — indeed most academics in the life sciences — "nobly" thought it ethically inappropriate and personally undignified to be motivated by financial largesse. The ultimate reward for scientific success is the Nobel prize — not filthy lucre — was the mantra! In the same vein, the opinion was widely proffered that the commercialization of academic research would not be in the best interests of maintaining free exchange of scientific information.

Cohen was therefore not especially excited when he heard from Reimers about the patent potential of recombinant DNA technology. Aside from the issues noted above (some of which may have motivated his initial lack of enthusiasm), Cohen knew that patented technologies didn't usually spin off substantive financial gains for some years, possibly not until the patent, typically awarded for 15 years, had expired. But he had a change of heart when Reimers educated him on the fundamental importance of patents, pointing out that "the commercial development of penicillin had been delayed for 11 years for want of a patent."[4]

Reimers also pointed out that in the event of a patent award, the licensing potential of recombinant DNA technology could be huge. "Patents and licenses arising from [inventions] are not only a mechanism for encouraging commercial development of basic science discoveries, but also potential generators of unrestricted funds that Stanford can then apply to support university research and education," Reimers stated.[5]

When corresponding with Bertram Rowland, the patent attorney engaged by Reimers, Cohen ultimately voiced his support to move forward with a patent application. "I can accept the view that it is more reasonable for any financial benefits derived from scientific research at a non-profit university with public funds to go to the university rather than be treated as a wind-fall profit to be enjoyed by profit-motivated businesses," he wrote. "I agreed to cooperate with Stanford for that reason."[6] Cohen duly alerted Reimers that Herb Boyer was a co-inventor and that he and the University of California at San Francisco would have to agree to any patent and/or licensing decisions.

In late 1974, Stanford approached Boyer and officials at the University of California at San Francisco. Both parties were willing to co-operate and Stanford was able to obtain approval from the American Cancer Society, the National Science Foundation and the NIH to administer the invention under Stanford's existing institutional Patent Agreement with the last-mentioned entity in time to file. The two universities agreed that Stanford would manage the licensing, that the university would deduct 15% of any gross income to offset administrative costs and out of pocket licensing expenses, and that remaining revenues would be shared equally by the two institutions. Cohen and Boyer were named as sole inventors on three patents filed by Stanford and UCSF. The first patent was granted to Stanford University on December 2, 1980. This and two subsequent patents are referred to as Cohen–Boyer recombinant DNA cloning patents.[7]

The May 1974 *PNAS* paper by Morrow *et al.* reporting the deployment of frog DNA in recombinant experiments listed six authors.

Rowland was concerned that the US Patent Office might question the restriction of inventorship to Cohen and Boyer, so he contacted John Morrow and Annie Chang at Stanford and Howard Goodman and Robert Helling at the University of California at San Francisco, requesting that they disclaim any inventorship. Morrow and Helling were not at all pleased when they learned that Stanford and the University of California were filing a patent exclusively in the names of Cohen and Boyer. They both refused to sign such a disclaimer. Helling expressed the opinion that "no university should be allowed to claim rights on a technology as fundamental and broadly applicable as recombinant DNA."[8] But such protests went nowhere. In the wake of this drama Helling threatened legal actions about having been left out of the patent and subsequent revenues. But in the final analysis he realized that would likely involve considerable personal financial expense ⋯ and eventually dropped the idea.

Some 10 years later, the NIH affirmed that universities in general should patent and license recombinant DNA technology. Meanwhile, in 1972 the General Electric Company filed patents on an organism that could break down multiple components of crude oil! That case went all the way to the Supreme Court, which voted 5–4 that a living, human-made microorganism is patentable subject matter.[9] The subsequent rapid and extensive commercialization of biotechnology in general, particularly in the private sector, radically altered the way in which business is conducted in the biomedical world. Perhaps the earliest and certainly one of the financially most successful private enterprises was the founding of the biotechology company Genentech by Boyer and others in San Francisco, a story well told by Sally Smith Hughes in her 2011 book *Genentech — The Beginnings of Biotech.*

Berg learned of the university's plan to apply for a patent on the basis of Cohen and Boyer's recombinant DNA work a few weeks before the meeting at Asilomar was to begin. He was especially outraged by that disclosure because of the damaging appearance that Stanford University scientists, who had been participants in calling for the "moratorium," were taking advantage of that lull for profit. His efforts to persuade the university to withdraw the patent application, or at least to assure the scientific community that a significant portion of the future royalties would be used to support research outside the university, perhaps through the auspices of the Research Corporation of America, fell on deaf ears. Further efforts through Reimers and the University Trustees failed to sway the course of events.

Berg attests that he has never gone out of his way to accrue financial reward from his many scientific contributions. But he heatedly drew Reimers' attention to the work from his own laboratory, proffering that "his own role and that of others in the development of recombinant DNA has been slighted."[10] Reimers was sympathetic. But he informed Berg that his early work was probably not patentable because it fell more in the category of an idea than a developed technology.

According to Stanford's technology-licensing practices, revenues received from patent royalties accrue one-third to the inventor, one-third to the academic department in which the work was done and one-third to the university. In this instance Cohen's share was distributed to the Departments of Medicine and Genetics. None came to the Department of Biochemistry or to anyone who contributed seminal ideas and/or experiments that made the Cohen–Boyer experiments possible. Predictably, these events did little to lighten existing tensions between Berg and others in the Department of Biochemistry — and Cohen. When in later years they documented the dramatic events that transpired in the late 1960s and early 1970s, Berg and his co-author Janet Mertz wrote:

> None of the members of the Berg, Kaiser, or Davis groups ever considered patenting the reagents or procedures that were used for recombining DNA *in vitro*. Neither had the scientists who discovered TdT, DNA polymerases, DNA ligases, exonucleases, and restriction enzymes ever sought patents for their efforts. Indeed, few, if any, of the discoveries, reagents, and methods that constitute the foundations of molecular biology were ever patented.
>
> While some academic institutions such as the University of Wisconsin — Madison had a long history of patenting inventions in the biological and biochemical sciences (e.g., vitamins and antibiotics), the sociology among most American life scientists prior to the 1970s was to eschew patents, believing that they would restrict the free flow of

information and reagents and impede the pace of discovery. However, that reticence disappeared in November, 1974 when Stanford University and the University of California at San Francisco jointly filed a United States patent application citing their respective faculty members Stanley Cohen and Herbert Boyer, as the sole inventors of the recombinant DNA technology. Their claims to commercial ownership of the techniques for cloning all possible DNAs, in all possible vectors, joined in all possible ways, in all possible organisms were dubious, presumptuous, and hubristic.[11]

In early October 1980 (a few weeks before he was informed that he had won the Nobel Prize in Chemistry), Berg gave an interview to Donald Stokes of the Stanford University News Service in which he expressed his displeasure at the commercialization of basic science:

> We shall all pay in the end. I am not against commercial backing of research. In some cases it is beneficial. What does appall me is the way it is seducing and subverting pure research, especially in the universities. ·· Research in recombinant DNA has opened up a whole new industrial sector. Some corporations are investing millions in it. Many people like to speculate that the next era of widespread discoveries and developments, comparable to what is happening in microelectronics, will be in the field of biochemistry. The money that is flowing is a significant factor in universities today. Yesterday they had never heard of us.
>
> The university says: "Why not? Why shouldn't we benefit? It's not as if we're pocketing the funds to build swimming pools, but rather to sponsor academic programs, to improve the teaching programs, to fund further research and so lead to more discoveries." I can understand all that. My worry about promoting the idea of patents is that it tends to bring along with it an element of secrecy. If you dangle the idea of patents and prospects of monetary reward in front of researchers it is bound to influence their thinking.[12]

In August 2001 Berg gave an intentionally sarcastic interview on this sensitive topic to the *San Francisco Chronicle*:

> When we spliced the profit gene into academic culture, we created a new organism — the recombinant university. We reprogrammed the

incentives that guide science. The rule in academe used to be "publish or perish." Now bioscientists have an alternative — "patent and profit."[13]

Berg was not the only Stanford opponent of the patents. Josh Lederberg also viewed patenting in universities as a deterrent to free and open scientific exchange. Arthur Kornberg voiced similar reservations. Had Berg, Lederberg and Kornberg been run-of-the-mill Stanford faculty members, Stanford would almost certainly have ignored their protests. But Lederberg and Kornberg were Nobel Laureates and Berg (who was soon to join their ranks) was then a highly visible media figure because of his prominent role in the recombinant DNA debates. "This was one of many situations occurring throughout the patenting process in which Stanford was compelled to weigh the benefits of commercialization against its political liabilities and to adjust activities in light of faculty dissension and broader social and political developments," Berg commented.[14]

Cohen initially waived his right to personal royalties, donating them to his *alma mater*, the University of Pennsylvania. He later decided to retain these financial proceeds, prompting Reimers to comment: "If anybody says that they understood that you waived royalties, I'm going to explain that you did, but now you don't — and refer them to you."[15]

Notes

1. Hughes SS. (2001) Making Dollars out of DNA: The First Major Patent in Biotechnology and the Commercialization of Molecular Biology, 1974–1980. *ISIS* **92**: 541–575, p. 545.
2. NR interview by Sally Hughes, Stanford's Office of Technology Licensing and the Cohen/Boyer Cloning Patents, 1997.
3. Hughes SS. (2001) Making Dollars out of DNA: The First Major Patent in Biotechnology and the Commercialization of Molecular Biology, 1974–1980. *ISIS* **92**: 541–575.
4. *Ibid.*
5. *Ibid.*
6. *Ibid.*
7. *Ibid.*

8. *Ibid.*
9. Kathy Ku, *Les Nouvelle*, p. 112, June 1983.
10. Hughes SS. (2001) Making Dollars out of DNA: The First Major Patent in Biotechnology and the Commercialization of Molecular Biology, 1974–1980. *ISIS* **92**: 541–575.
11. Berg P, Mertz J. (2011) Personal Reflections on the Origin and Emergence of Recombinant DNA Technology. *Genetics* **184**: 9–17.
12. Draft of article by Donald Stokes, dated October 7, 1980. Reproduced with permission from the Department of Special Collections, Stanford University Libraries.
13. *San Francisco Chronicle*, August 13, 2001.
14. Hughes SS. (2001) Making Dollars out of DNA: The First Major Patent in Biotechnology and the Commercialization of Molecular Biology, 1974–1980. *ISIS* **92**: 541–575.
15. Interview of Niels Reimers by Sally Hughes, Regional Oral History Office of the Bancroft Library at the University of California, Berkeley, 1997.

CHAPTER TWENTY SEVEN

Life Goes On

But we missed the existence of introns

With the full ripening of recombinant DNA technology and gene cloning, the field of eukaryotic molecular biology literally exploded. Securely ensconced in his Stanford laboratory once again, Berg and his research group continued to plumb the depths of SV40 biology, initially making concerted efforts to map genes in the viral genome. Other than the knowledge that about half of the SV40 genes are expressed early and the rest late, the precise location and organization of these so-called early and late genes was unknown.

Some temperature-sensitive mutants were then available for mapping studies. However, Berg prompted Thomas (Tom) Shenk, a postdoctoral fellow in his group, and John Carbon, on leave from the University of California at Santa Barbara, to launch a major effort toward generating new mutant strains, especially those carrying deletion mutations — which have the obvious virtue of being non-revertible.[1] More importantly, their location in the SV40 genome could be identified by restriction enzyme fragmentation patterns and by electron microscopy.

Such mutants could be recovered after single cleavages of the DNA with site-specific restriction enzymes, or after limited random cleavages throughout the DNA, as John Morrow had done for his PhD thesis. By relying on transfected cells to rejoin DNA ends generated by cleavage *in vitro*, two classes of mutant populations were obtained. One could be propagated as efficiently as the original wild-type virus, regardless of the fact that the various mutant strains bore deletions of varying size. The

Berg at work with postdoctoral fellows Tom Shenk and Steve Goff (*left*).
(Courtesy of Paul Berg.)

other class could not be propagated without co-infection with a virus that supplied the disrupted function(s). The location and size of the deletions of both classes of mutants could be readily determined using the technologies mentioned above.

A particular class of viable deletion mutants was found to lack variable lengths of DNA at a site that was suspected to be within the coding region of an essential gene, the so-called large T gene essential for viral DNA replication and known to be expressed early in SV40 infection. Yet these viruses grew normally, were capable of converting normal mouse cells into cancerous cells and produced normal-sized large T protein. How could this be? Shenk, Carbon and Berg published their observations in May 1976, but at this early stage it was not compelling to consider the possibility of RNA splicing as an explanation.

In July, August and September 1977, Richard (Rich) Roberts and his colleagues published three seminal papers in the high-profile journal *Cell*. The July paper reported the presence of a predominant 5'-undecanucleotide (11 nucleotides) in the late class of RNAs. In the discussion section of their paper, Richards *et al.* offered the suggestion that "the presence or absence of this undecanucleotide is associated with the control mechanism that defines the switch from early to late transcription." They continued this line of conjecture with the statement: "*A simple model by which this might be accomplished could involve the specific recognition of all or part of this common sequence present at many sites within a large nuclear tran-script, followed by processing to give the discrete cytoplasmic mRNA species* [author's italics]."

The September 1977 issue of *Cell* contained five articles concerned with this general topic, including one by Roberts *et al.* and one by Daniel Klessing, another independent investigator at Cold Spring Harbor. When concluding their discussion, Roberts stated that the results presented in this clutch of papers *"are not directly consistent with any mechanism previously suggested for the biosynthesis of mRNA in eucaryotic cells. They imply that an alternate scheme must exist for Ad2-SV40, and perhaps for eukaryotic mRNA in general* [author's italics]."

Essentially similar conclusions were independently arrived at in the same time frame by Philip (Phil) Sharp and his colleagues at MIT. Both research groups stunned a large audience at the 1997 Cold Spring Harbor Symposium with their presentations. However, publication of these presentations did not transpire until 1978. Both articles explicitly or implicitly referred to RNA splicing. Sharp and Roberts were awarded Nobel prizes in 1993.

Berg retained his interest in the biology of SV40 virus until the end of the 1970s. As mentioned earlier, in 1977 he spent a brief sabbatical at the Imperial Cancer Research Fund (ICRF) in London collaborating with a British virologist, Alan Smith. The pair observed that while infections with wild-type SV40 virus rendered a second protein, related to but smaller than large T (little T as it came to be known), this protein was missing in cells infected with the Shenk–Carbon deletion mutants. Remarkably, while Berg was at the ICRF, Lionel Crawford, an English scientist from that institution crossed the Atlantic in the opposite direction and spent a mini-sabbatical in Berg's Stanford laboratory. Crawford, together with Stanford postdoctoral fellow Charles Cole, made the identical observation.

Both groups realized that while the deletion in the middle of the large T gene did not affect the size or function of large T protein, it eliminated the formation of little T, suggesting that the small T protein is normally partially encoded by sequences in an intron. As it turned out, the primary RNA transcript can undergo either of two alternative splicing events, one yielding large T protein and the other little T protein.[2]

Neither Berg (who was then in London) nor Shenk or Carbon attended the 1977 Cold Spring Harbor Symposium. But the news spread rapidly, and when they heard and/or read about this news it became evident that the viable deletions in the SV40 genome that the pair were studying were likely located in an intron resident in the large T gene — and that the Berg laboratory had essentially lost an opportunity to claim priority for the discovery of introns and mRNA slicing. Berg still champs at the proverbial bit at having "missed out" on this fundamental discovery! Who wouldn't?

Missing the discovery of introns and the existence of splicing as another step in the genesis of mRNA was an embarrassing and somewhat disheartening experience for Berg. But his despondency at being "scooped" by the Sharp and Roberts laboratories was somewhat ameliorated by winning a bet with Francis Crick on the biological significance of introns and splicing as important properties of all eukaryotic genes. To wit, after Berg presented the discovery of introns and splicing at a seminar at the Salk Institute, Crick took exception to his prediction of alternative splicing as a widespread feature of the expression of mammalian genes. Instead, he proffered that such a biological device was a property unique to viral genomes, which evolved to cope with the constraints of packaging large quantities of DNA into mature virions. He predicted that other eukaryotes would be shown to duplicate or expand the number of resident genes in their genomes rather than utilize a single gene in multiple ways. Both being ardent wine connoisseurs, Berg and Crick bet a case of fine wines on that difference of opinion.

After several years, as examples of genes with multiple introns became commonplace and the relevance of alternate splicing to generate different proteins from a single gene became indisputable, Berg sent Crick a reprint of a paper documenting the formation of two different brain proteins with entirely different functions, which were derived from the same gene. Shortly thereafter he had cause to drop an oblique thank-you note to Crick.

> Dear Francis,
>
> One case of wine arrived at my doorstep and a call to the wine store confirmed my assumption that it was from you. ....
> ....

Please give my regards to Odile. I hope you are well and if you -----
come this way there'll be plenty of wine to share.³

The original goal of using SV40 as a means for introducing new genes
into mammalian cells was ultimately realized when Steve Goff, a graduate
student with Berg, constructed a variety of recombinant molecules
containing foreign DNA. These included recombinants with a segment
of λ phage DNA containing several protein coding genes — the *E. coli*
thymidine kinase gene and a yeast tyrosine-specific tRNA gene. These
recombinants could not be propagated alone, but proliferated normally
in the presence of a helper virus that provided the function(s) of the missing region of DNA.

Richard (Rich) Mulligan, another graduate student in the Berg camp,
succeeded in replacing the gene encoding the SV40 major viral capsid
protein with the coding sequences for rabbit α- and β-globin proteins or
with those in a cDNA from the mammalian *DHFR* gene, and demonstrated the presence of appropriate mRNAs and their corresponding
proteins after infecting monkey kidney cells with the *DHFR* cDNA.
These experiments provided the first demonstration of a mammalian
gene from one species functioning in another eukaryotic organism.⁴
Mulligan also generated recombinants of SV40 DNA containing several
bacterial genes — *got* and *neo* — each of which encodes a property that
renders transformed cells resistant to antibiotics. This strategy soon
became widely used for introducing genes into mammalian cells.

Around that time, the Berg laboratory constructed several novel
cloning vectors containing genes for selection in both *E. coli* and in mammalian cells. Vectors were also designed to replicate in both bacterial and
animal cells. Both of these plasmid-like vectors became widely used to
facilitate the introduction and maintenance of foreign DNA into both
bacterial and mammalian cells. In 1981, a talented Japanese postdoctoral
fellow, Hiroto Okayama, made a particularly significant technological
advance by devising a strategy for cloning full-length DNA copies of
mRNA (so-called expression cloning) and for facilitating functional com-
plementation of mutant phenotypes. The Okayama–Berg cloning vector
became widely used in academic studies and in the burgeoning biotechnology industry.

Still stung from Berg's outburst surrounding the patents filed by Cohen and Boyer, Neils Reimers of the Stanford patent office promptly contacted Berg. "I don't know whether there is any commercial potential in what you have done, or if there are novel techniques that you used that would be patentable," Reimers wrote. "If there are, you might want to consider filling out a disclosure."[5] Berg never took Reimers up on his offer.

****

Notwithstanding his crowded scientific, administrative and public policy agendas, Berg at one time served as a Non-resident Fellow of the Salk Institute in La Jolla, California. When the germ of the idea for creating a research institute in the opulent setting of La Jolla California emerged from Jonas Salk, who discussed the concept with a number of prominent scientific colleagues, including Renato Dulbecco, Melvin Cohen, Ed Lennox and Leslie Orgel. Salk's original plan was to build an institute that in addition to basic scientists included prominent intellectuals in the fine arts — a true Renaissance experiment!

Sometime in the early 1960s Salk interviewed Berg to determine his level of interest in joining such an enterprise. Berg was non-committal. Around that time Berg also met with the eminent physicist-turned-biologist, Leo Szilard, who had been quietly working behind the scenes with Salk ever since the idea of an institute in La Jolla was first floated. The meetings were ostensibly to determine whether Berg was "suitable" to be included as one of the selected Resident Fellows. Nothing came of the two interviews. Many years later, Berg learned from Mel Cohen's wife, Suzanne Bourgeois (who was then writing a history of the Salk Institute entitled *Genesis of the Salk Institute — The Epic of its Founders*, scheduled to be released in late 2013), that in the course of her research she had uncovered a letter from Szilard to Salk indicating that Berg was not suitable to be a Fellow in an institute that aspired to an intellectual mix of scientists, philosophers, artists and other intellectuals — because he was "just a biochemist!"

To complement the activities of the Resident Fellows, Salk envisaged a larger group that he referred to as Non-Resident Fellows. Such

Jonas Salk (*left*) with Salk Resident Fellow Renato Dulbecco.
(http://www.salk.edu/insidesalk/images/0712/dulbecco-home)

individuals (also to be scrupulously selected) were expected to visit the institute on a regular basis, generally for a month or so when the annual summer meetings with the Salk research groups took place. The primary purposes of these summer visits were to promote dialogue with individual investigators, lead discussions on future scientific directions, and advise the institute director and the Resident Fellows on internal issues such as promotions and recruitment of new appointees.

Berg was invited to become a Non-Resident Salk Fellow for a 10-year term, beginning in 1978. Having already spent a year at the Salk for his providential sabbatical in the late 1960s, during which he participated in these summer activities, he was sufficiently intrigued — and agreed to serve in this capacity.[6] Other Non-Resident Fellows included Francis Crick, Salvador Luria, Gerald (Gerry) Edelman, Leo Szilard and Steve Kuffler (a pre-eminent Hungarian-American). Berg especially treasured interactions with the institute's younger scientists, notably Tony Hunter, Inder Verma, Ron Evans, Fred Gage and Jeffrey Lon (all of whom progressed to notable scientific careers), as well as the many postdoctoral fellows that populated the research laboratories.

****

"When recombinant DNA technology exploded in the 1980s, some of us quickly recognized its potential for revolutionizing medicine in profound ways,"[6] Berg stated. Since Stanford scientists had played such a prominent part in the 'genetic revolution,' we reasoned, why shouldn't Stanford be a leader in promoting such possibilities?" Another strongly motivating factor accrued from the long-standing observation that many of the Stanford clinicians were not yet caught up in what had transpired toward transforming genetics to a molecular science. So, his hands not full enough for his liking, Berg began agitating for a new building to house a center for genetic medicine.

To jump-start the initiative, he and a small committee organized a one-day conference on medical/molecular genetics that featured some of the leading lights in the field of molecular medicine. The group included NIH director Don Fredrickson (who was soon to become the president of the Howard Hughes Medical Institute). Fredrickson was of the firm conviction that bringing basic scientists and clinically oriented scientists together was essential to the future of medicine and strongly encouraged Berg and his colleagues to move forward with concrete plans.

Berg learned that Arnold Beckman (the founder of a highly financially successful business entity called Beckman Instruments and a previous donor to numerous scientific and other academic entities around the country) might consider supporting such an initiative, this despite the fact that Beckman had no known previous ties to Stanford University. "So we invited him to spend a few days with us," Berg commented. "He came to Stanford and met with the key players. But after he departed it proved to be extremely difficult to contact the man. From University president Donald (Don) Kennedy down, no one could get through to Arnold Beckman!"[7]

In the interim, Stanford president Kennedy met with Berg and informed him that the university was going to drop the entire project unless Berg agreed to serve as scientific director, a move that Kennedy thought would put teeth into the project and enhance its selling power. After much soul searching Berg agreed to assume that responsibility — provided that Kennedy acceded to two conditions. One was that the entire Department of Biochemistry would move into the new center. "I wasn't about to move and leave my colleagues of 20 years behind in our original facilities."[8] The other condition was that construction would commence before all required funding was in hand. Kennedy acceded to both requests.

The renewed vigor prompted Berg to more aggressive attempts to contact Beckman — with eventual success. He was now in a position to relate concrete progress, and as he hoped would be the case, he was invited to meet with Beckman once more. Kennedy, Arthur Kornberg, John Ford (an experienced university fund raiser) and Berg traveled to Beckman's Southern California office. "We spent 30 to 40 minutes describing our ambitions and plans." Berg related, "during which I became concerned because at times Beckman seemed not to be paying much attention."[9] Regardless of whether or not he was in fact fully attentive to the group's presentation, Beckman acceded to providing Stanford with $12.5 million, plus an additional $3.5 million for a crystallography facility in the adjoining Simon Fairchild building. In fact, Beckman was enthusiastic enough about the plans outlined that he persuaded a friend to provide $1 million to the project. Soon after Don Fredrickson assumed the presidency of the Howard Hughes Medical Institute (HHMI), he agreed to provide approximately $15 million for space in the projected new building to support 12 HHMI investigators for a decade.

In addition to HHMI faculty, this initiative spawned two new academic departments at Stanford medical school — a Department of Molecular and Cellular Physiology and a Department of Developmental Biology — the first of its kind in US medical schools. The Mable and Arnold Beckman Center for Molecular and Genetic Medicine opened its doors in the spring of 1989.

This opening also marked the beginning of the end of Berg's long and productive scientific career. "Managing the creation of the Beckman Center enterprise was time-consuming and my research program suffered," he related. "The field of molecular biology was expanding enormously and I simply wasn't able to keep up with it at a level to which I was accustomed. Additionally, I was convinced that I was no longer attracting the quality of students and fellows that had joined my laboratory during the halcyon days in the 1960s–1980s. So the thought of having to provide funding for what I believed was no longer an exciting research program was discouraging." Berg dabbled briefly in the biochemistry and molecular biology of genetic recombination. But in 2000, at age 74, he closed his research laboratory and passed the directorship of the

Arnold Beckman. (www.beckman-foundation.com)

Beckman Center to his colleague Lucy Shapiro, who had been recruited to lead the new Department of Developmental Biology.

****

Young readers may find it surprising that "moonlighting" in the private sector as a paid consultant, advisor, or in any other professional capacity that may yield financial gain was seriously frowned upon until the widespread commercialization of recombinant DNA technology. Aside from his principled disdain for scientists so involved, Berg believed that it would be particularly unethical if not overtly hypocritical for him to indulge in such activities while in the role of spokesperson for the safe use of recombinant DNA technology. Nonetheless, as the controversy simmered down at the end of the '70s, Berg, Kornberg and Charles Yanofsky were enticed to contemplate joining a biotech startup company by a Stanford professor, which was being funded by a local venture capital firm. Before they signed on to that venture, however, a prominent entrepreneur, Alejandro Zaffaroni entered their lives.

Born in Uruguay, Zaffaroni received a PhD degree in biochemistry from the University of Rochester in 1949. While at Rochester and in the years immediately following, Zaffaroni cultivated a keen business sense and became interested in the innovative application of cutting edge pharmacology in the private sector. In due course he lent his skills to a small Mexican chemical company that specialized in steroid research. Under

Zaffaroni's skilled management, Syntex Corporation (as the company was called) became a major multinational Palo Alto-based pharmaceutical company that among other things pioneered the development of oral contraceptives.

Eager to establish more effective and efficient ways of regulating the delivery of drugs and optimize their pharmacologic efficacy, Zaffaroni founded a new company in Palo Alto California called ALZA, which he soon transformed into the world's leading drug-delivery technology company. ALZA's portfolio included financial blockbusters such as contraceptive devices and the *NicoDerm* patch for smoking cessation. When Zaffaroni learned that Stanford faculty of the ilk of Kornberg, Berg and Yanofsky were considering joining the new biotech company, he interceded with an offer of his own.

Alejandro (Alex) Zaffaroni.

Kornberg introduced Zafferoni to Berg and Yanofsky, who quickly identified his keen appreciation for melding free enterprise with high quality science. Zaffaroni invited the trio to consider joining him as members of the Scientific Advisory Board of his new venture. He named the new company DNAX, an entity with the initial goal of developing genetically engineered antibodies. The venture prospered and within 18 months was acquired by the Schering–Plough pharmaceutical company.

"Alex was a visionary individual with an enormous respect for basic science," Berg commented. "At one point he was considering making an offer to Stanford to acquire shares in DNAX in return for a license to

Henry Kaplan's (former chairman of the Department of Radiology at Stanford and then a powerful voice in medical school affairs) collection of human monoclonal antibodies." But Berg's concerns about the corruption of academia by industry and its promise of financial gain led him to oppose the arrangement. "Alex's response was unequivocal," Berg remarked. "If you're opposed to it Paul, we won't do it," he said, a rejoinder that deeply impressed Berg. "Subsequently, other opportunities that made more sense from a business point of view were discarded by Alex because they might be detrimental to the science at DNAX," Berg stated.[10] With advice from Berg, Yanofsky and Kornberg, and under the scientific direction of a bright former postdoctoral fellow in Kornberg's laboratory, Kenichi Arai, DNAX, now a subsidiary of Schering–Plough, became pre-eminent in the cloning of cytokine cDNAs and genes, and their receptors.

In Berg's opinion, the basic science infrastructure at DNAX at that time rivaled that of the famous Basel Institute of Immunology in Switzerland. "We had excellent academic people in the company," he related. "A lot of young folks made their scientific reputations there. Zaffaroni operated on the philosophy that it was the company's responsibility to encourage scientists to publish their research and to present their findings at meetings. Essentially, if DNAX lawyers were required to file a patent on a Saturday night so that one of the scientists could make a presentation at a scientific meeting the following Monday — that's the way it was done."[11]

Berg's association with Zaffaroni at DNAX blossomed into participation in a second venture, Affymetrix, which led to the development of DNA microarrays and their widespread adoption for genome scanning and monitoring gene expression. Beginning in the latter part of the 1980s, Berg spent 18 years as a member of the Affymetrix Board of Directors, a period he regards as having been intellectually, scientifically — and financially — rewarding.

Some years later, Berg met and became friendly with George Shultz, who had served as Secretary of Labor, of the Treasury and of State under multiple presidents. Schultz had retired from government and was (he still is) a Distinguished Fellow at the Hoover Institution, a think tank on the Stanford campus. Soon thereafter, Berg (whose name was increasingly becoming associated with successful private companies) was invited to become a member of the Board of Directors of Gilead Sciences, a Bay Area company then focused on anti-viral therapies, which ultimately rose to become the premier developer of antiretroviral drugs specifically for the treatment of HIV infections. Berg served as a member of the board for 13 years and currently continues this relationship as a scientific adviser.

Regardless of significant financial reward for all these efforts, Berg and his wife Millie lead modest material lives. Their home on the Stanford campus, purchased when they moved to the Bay Area in 1959, suits their needs admirably and the couple has no plans to change that comfortable situation. They are not avid travelers and their son John lives in the area, so there are few if any demands on family travel. Annual trips to New York to take in good theater, art museums and galleries, and visit with his brother and his family have become a regular event. Trips to London, Paris, and other continental centers are now less frequent than in years gone by. The San Francisco Bay area now provides all the intellectual, artistic and dining opportunities the couple can manage. Hence, in their later years the Bergs have welcomed the opportunity to benefit others as generous philanthropists.

Notes

1. ECF interview with Paul Berg, June 201.
2. E-mail from Paul Berg, February 1, 2012.
3. Letter from Paul Berg to Francis Crick, February 13, 1986. Reproduced with permission from the Department of Special Collections, Stanford University Libraries.
4. Mulligan RC, Howard BH, Berg P. (1979) Synthesis of rabbit β-globin in cultured monkey kidney cells following infection with a SV40 β-globin recombinant genome. *Nature* **277**: 108–114.
5. Letter from Niels Reimer to Paul Berg, November 3, 1978. Reproduced with permission from the Department of Special Collections, Stanford University Libraries.
6. ECF interview with Paul Berg, February 2012.

7.   ECF interview with Paul Berg, August 2012.
8.   *Ibid.*
9.   *Ibid.*
10. *Ibid.*
11. *Ibid.*

CHAPTER TWENTY EIGHT

The "Retirement" Years

I expect I'll be dragged into the fund raising
for that project

Resigning the directorship of the Beckman Institute and closing his research laboratory did not leave the troublesome vacuum in Berg's professional life that many retired academics dread. At the time of this writing he has continued serving on the Gilead Board of Directors and as director of science at Affymetrix in emeritus capacities. He eagerly follows scientific progress at both companies, frequently stopping by the research laboratories to chat with the scientists — especially the younger members. He also conscientiously reads the premier science journals to keep up with progress in the biological sciences, and was readily available for consultative advice to Philip Pizzo, Dean of the Stanford Medical School from 2001–2012.

Soon after taking office as dean, Pizzo (whose uncensored admiration for Berg has already been mentioned, see Chapter 2) turned his attention to long-standing limitations in the medical school's teaching activities. The US Department of Education recognizes the Liaison Committee on Medical Education (LCME) for accreditation of medical education leading to the MD degree. Hence, in order to retain accreditation, American medical schools are required to pass muster following periodic formal inspection. These inspections are treated seriously by the LCME. They require months of preparation by schools due for inspection and unfavorable reports can lead to probationary status until deficiencies are corrected.

A report following an LCME visit to Stanford Medical School shortly before Dean Pizzo took office in 2001 was notably critical of the outmoded and crowded state of educational, learning and recreational facilities for medical students. Accordingly, when he stepped into the Dean's office Pizzo categorically stated: "We have to do something innovative to rectify this situation!"

Berg was "invited" to join a task force to identify and implement an acceptable strategy. Lengthy deliberations led to the decision to erect a new Center for Learning and Knowledge at Stanford University Medical School. A fund-raising campaign was launched with the goal of raising $51 million to erect a new building to house an innovative and forward-looking educational center.

With an estimated wealth of US$26.0 billion in mid-2011, Sir Ka-Shing Li is claimed to be the richest person of Asian descent in the world. A noted philanthropist, in 2006, Li pledged to donate one-third of his entire fortune to charity and philanthropic projects throughout the world — a pledge estimated at over US$10 billion.[1] Having earlier turned down a request from Stanford to launch the Beckman Center, the Li Ka-Shing Foundation assumed almost half the proposed cost of the new education entity. The architecturally splendid Li Ka-Shing Center for Learning and Knowledge (LKSC) was formally dedicated in 2010.[2]

The Li Ka-Shing Center at night. (http://lksc.stanford.edu)

The Paul and Mildred Berg Hall at the Li Ka-Shing Center for Learning and Knowledge.
(http://conferencecenter.stanford.edu/)

The five-story building is endowed with a diverse array of sophisti-
cated technologies — including one of the largest and most advanced
simulation facilities in the US as well as a high-end video capture system
that represents the latest in medical education.[3] The second level of the
LKSC houses a large conference center entered via staircases at the front
and back of the building. The Paul and Mildred Berg Hall in the conference center recognizes a major gift from Paul and Millie Berg.

****

In late March 1979, Berg presented three invited public lectures
entitled *Dissecting and Reconstructing the Molecules of Heredity: An
Expanding Chemical Frontier*, to help inaugurate an annual Distinguished
Lecture Series at the University of Pittsburgh. Berg being a decidedly
articulate speaker with a knack for pitching talks at a level appropriate to
any audience, his lectures (which were open to the general public) were a
resounding success.

Not long after returning to Stanford, Berg received the customary
warmly written thanks from the University of Pittsburgh. He was additionally invited to reproduce the content of his lectures in a published
format. Berg acceded to the request — in principle! Time passed, yielding
periodic gentle but insistent reminders of his promise. Eventually, to
everyone's surprise (most certainly to the University of Pittsburgh),
despite a long laundry list of "things to do," Berg offered to write a full-
scale textbook — with Maxine Singer as his co-author. "She and I had

from time to time talked about writing a monograph about the science and public policy aspects of recombinant DNA but never got so far as to begin. Maxine is a marvelous writer and a keen student of the subject," Berg related.[4] The Dean of Graduate Studies at the University of Pittsburgh was unsurprisingly thrilled at the prospect.

Slap in the middle of this writing, Berg formally assumed the directorship of the new Arnold Beckman Center for Genetic and Molecular Medicine. "I was planning the construction of the building and the recruiting of 15–20 new faculty members, not to mention the woes and challenges that always accompany such enterprises. Maxine was busy with her own research and soon after we began writing she was invited to become president of the Carnegie Institution for Science in Washington, DC."[5]

Maxine Singer first met Berg as a graduate student at Yale when he visited the university shortly after completing his postdoctoral stint in Copenhagen. An outstanding molecular biologist and an astute administrator in her own right, Singer became head of the Laboratory of Biochemistry in the Cancer Institute at the NIH before assuming the presidency of the Carnegie Institution of Science in 1988. She retired from that position in 2002 after winning several notable awards, including the National Medal of Science and the National Academy of Sciences' Public Service Award. Berg's high regard for Singer's academic talents and "good sense" accounted for his invitation to her to join the Asilomar Organizing Committee in 1973. Not surprisingly, when struggling to launch a significant textbook on modern genetics and molecular biology, Berg sought Singer's assistance as co-author of the intended work.

"The best thing about writing a book together when the authors are separated by 3,000 miles is that you don't interfere much with each other's writing," Singer commented. "We divided chapters. He worked at his pace; I worked at mine. Our writing styles were similar enough that we didn't have to write and read more than a couple of drafts. But I didn't agree with everything he wanted to do — and we had a few knockdown drag-outs while writing! One has to be pushy with Paul sometimes," Singer interjected. "He loves to talk and it's sometimes difficult to get a word in edgeways. That's a well-known mannerism among Brooklynites. If one understands Brooklynites one can understand Paul Berg! One also has to get used to the notion of 'I'm always right!'" Singer relates that Berg once had a sign on his office door that read: *Once I thought I was wrong. But I was mistaken!* "That would be a great title for a book about Paul," she laughingly suggested.[6] (Being a fly on the wall when Berg and Singer were discussing features of the book they agreed on — and especially those they sometimes disagreed on, would have been a rare treat.) In due course, taking time away from their respective offices and the attendant pressures of other obligations became a necessity. About three weeks of several summers were spent at the Institute for Advanced Studies at Princeton University, an arrangement based largely on Berg's friendship with Harry Woolf, director of the institute, who extended an open invitation to the pair to spend summers in Princeton. "We didn't have to pay for a thing, including food," Berg stated.[7] Singer and Berg also journeyed to Woods Hole, for many years a favorite summer retreat for scientists. Berg and Singer occupied the place during winter. "Accommodations were Spartan and the food was nothing to write home about. But the library there was superb," Berg stated.[8]

In 1991, 22 years after delivering the set of invited lectures at the University of Pittsburgh, a tome of over 900 pages entitled *Genes and Genomes — A Changing Perspective* emerged.[9] Lavishly illustrated in color, the textbook is divided into four main sections that address the molecular basis of heredity, the emergence of recombinant DNA, expression and regulation of eukaryotic genes, and a section devoted to understanding and manipulating biological systems. The book was well received. "A superb textbook suitable for college seniors and first-year students," the journal *SCIENCE* reported, while *Nature* wrote: "Topics are treated in greater depth and with more insight than in any other text currently available."

Encouraged by this warm reception, Berg and Singer elected to follow *Genes and Genomes* with a more abbreviated work called *Dealing With Genes — The Language of Heredity*, published in 1992. "Throughout the first textbook project we discussed the importance of telling the story about modern genetics and molecular biology to a broader readership,"

Berg related.[10] Reviewers of the textbook agreed. In the preface to *Genes and Genomes,* Singer and Berg communicated their interest in rendering a version of the book that "told the story to a broader group," including interested lay people. "Biology and genetics, even molecular genetics, in contrast to modern physics, do not depend on abstract concepts, nor does their understanding require a mastery of mathematics," they wrote.[11] Reviewers' praise for *Genes and Genomes* prompted the authors to render yet another work, *Exploring Genetic Mechanisms*, dedicated to Harland Wood and Arthur Kornberg (who both greatly influenced Berg's career), and to Joseph Fruton and Leon Heppel (important mentors to Singer); the preface to this tome states:

> Our earlier textbook, *Genes and Genomes*, used well-studied examples to show how modern experimental techniques provide rich molecular detail about genetics. That book and other textbooks with similar intentions stress the unity of genetic processes among different organisms. ·········· Molecular analysis has only begun to reveal the varied genetic tactics that account for the diversity of organismal form, habitat, behavior, and function. ······························· This book aims to introduce to students and scientists how such complexity is being analyzed.[12]

Significantly, this time the authors substantially lightened their writing load by inviting a cadre of 15 high-profile molecular biologists and geneticists to contribute all but two chapters, which they wrote themselves. Published in 1997, *Exploring Genetic Mechanisms* drew numerous complimentary reviews.

But Berg and Singer were not done with writing books! Their next literary adventure was a switch from textbooks to a biography of the famous American geneticist George Beadle, who together with Edward (Ed) Tatum and Josh Lederberg garnered Nobel prizes for defining the role of genes in regulating biochemical events in cells, a discovery that led to the famous aphorism, "one gene, one enzyme." This literary contribution by Berg and Singer (Singer being first author of all three previous

Paul Berg and Maxine Singer at the time they were writing books together. (Courtesy of Paul Berg.)

books) was seven years in the writing. The book had an interesting origin.

Sometime in the mid- to late 1990s, Berg was invited to lunch in Arthur Kornberg's office in the Stanford Department of Biochemistry. The lunch was in honor of Horace Albert "Nook" Barker, a distinguished scientist, then a retired faculty member from the University of California at Berkeley. In appreciation of his pioneering use of radioactive carbon-14 tracers in biochemical research in the mid-1940s, and possibly even more so for his discovery of vitamin B12 in the 1950s, Barker was recognized with the Medal of Honor from President Lyndon Johnson in 1988.

Berg was scheduled to meet with a group of graduate students imme- diately following the luncheon, but the visit with Barker and Kornberg lasted longer than he had planned. When he finally met with his students he offered his apologies and briefly explained why he had been delayed. "When I mentioned the name Nook Barker I saw nothing but blank faces," Berg commented. "So I explained who he was and also mentioned the names of a number of other past prominent biochemists and geneticists. The students hadn't heard of any of them."[13] Expressing his frank amazement at such ignorance, Berg chided the class. "You guys are aiming to become biochemists and you know nothing about the major figures in the discipline?" he asked incredulously.

The level of ignorance exhibited by his students continued to bother Berg, who ultimately decided to offer a course on the history of science that would focus primarily on two questions: "How did we get to the genetic code?" and "Who were the people and what were the key events that lead ultimately to that?" In so doing he decided to trace genetics and molecular biology from the beginning of the 20th century, selecting (among others) papers published by the likes of William Garrod, George Beadle, Edward Tatum, Oswald Avery, Colin MacLeod and Maclyn McCarty, Sydney Brenner, Charles Yanofsky, Francis Crick, microbiologist M. H. Dawson, Boris Ephrussi, George Gamow, Al Hershy, Rosalind Franklin, Linus Pauling and Elliot Volkin and Larry Astrachan.

The class convened once a week and was given papers from the literature to read for discussion. At the end of the quarter, the students unanimously voted the experience the best course they had had since coming to Stanford, no doubt a tribute to the intrinsic importance of the topic, but most certainly also to Berg's engaging style as a lecturer! Without question the long-standing and widespread complaint that graduate students in the life sciences have little if any appreciation of history was thoroughly debunked by Berg that semester.

George Beadle ("Beets" to his friends), one of the scientists singled out for attention during the course, provoked particular interest in the class and stirred Berg, now an ascending polymath, to the idea of documenting Beadle's biography. "The book offered us the opportunity to write a history of 20th century genetics," Singer stated. "Early in his professional life Beadle worked in Thomas Hunt Morgan's laboratory at Caltech and he trained a good number of future molecular biologists. And it was totally new to both of us."[14]

A quick visit to Beadle's archives at the California Institute of Technology (Caltech), where he worked from 1931–1937 and again from

1954–1960, persuaded Berg and Singer that "we had enough material for a good book," a view emphatically reinforced, certainly in Berg's mind, by Beadle's sojourn at Stanford from 1937–1954, during which time he worked with Tatum on the mold *Neurospora,* work that was honored with Nobel prizes in 1958. Berg and Singer also visited Wahoo, Nebraska (population 4,508 at the 2010 census) where Beadle was born and raised on a farm operated by his parents. He might himself have become a farmer if one of his teachers at school had not directed his obvious intellectual talent to science and persuaded him to attend the College of Agriculture in Lincoln, Nebraska.

Wahoo turned out to be a treasure trove of Beadle memorabilia. "If not for George Beadle I would never have visited Nebraska," Singer joked. Berg and Singer completed the book, their last as co-authors, in 2003, when Cold Spring Harbor Laboratory Press — to wide acclaim, published *George Beadle — An Uncommon Farmer.* In an essay entitled "A Scientific Entreprenuer" for *American Scientist,* Bruno Strasser referred to

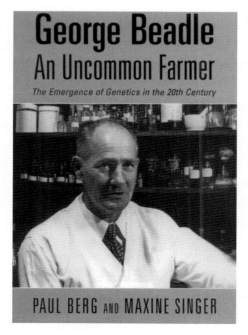

Published in 2003. (Courtesy of Cold Spring Harbor Laboratory Press.)

the book as "a meticulously investigated, historically contextualized and finely written biography of one of the central figures of 20th century genetics."[15] In 2003, Berg and Singer additionally co-authored an article with Norman Horowitz, Josh Lederberg, John Doebley and James Crow (all renowned geneticists) in celebration of the centennial of Beadle's birth.[16]

When asked by the author (very much with tongue in cheek) whether he contemplated any further book writing, Berg was initially equivocal. In the not too distant past he had seriously contemplated writing a history of the discovery of introns and RNA splicing, a story he is intimately familiar with based on the near miss by his own laboratory related earlier. "I started writing, but soon found that I no longer have the stamina for this sort of effort," he related at age 86. "And had I carried on I would have expected that I would have had to do interviews — and travel is no longer comfortable."[17]

Berg may have lost his appetite for full-length literary contributions that require in-depth research. But his penchant for writing (an occupation at which, as already mentioned, he is most adept) and expressing his views on all manner of issues, has not prevented him from composing several autobiographical articles as well as a piece about his brief involvement as chairman of the National Advisory Committee Human Genome Project entitled *Origins of the Human Genome Project: Why Sequence the Human Genome When 96% of it is Junk?*[18] Then too, Berg (sometimes with Singer) has contributed an extensive collection of commentaries and opinion pieces to high profile journals, including *Nature*, *PNAS* and *SCIENCE* — not to mention the many letters of protest, admonition, indignation or congratulation that continue to flow off his pen. There is little question about Berg's literary skills and his penchant for exercising them.

Notes

1. http://en.wikipedia.org/wiki/Li_Ka-shing#Philanthropy
2. http://lksc.stanford.edu
3. http://med.stanford.edu/featuredtopics/2010/lksc.html

4. ECF interview with Paul Berg, August 2012.
5. *Ibid.*
6. ECF interview with Maxine Singer, February 2012.
7. ECF interview with Paul Berg, August 2012.
8. *Ibid.*
9. Singer M, Berg P. (1991) Genes and Genomes, University Science Books.
10. ECF interview with Paul Berg, February 2012.
11. Singer M, Berg P. (1991) *Genes and Genomes*, University Science Books.
12. Singer M, Berg P. (1997) *Exploring Genetic Mechanisms*, University Science Books; http://en.wikipedia.org/wiki/Horace Barker
13. ECF interview with Paul Berg, August 2012.
14. ECF interview with Maxine Singer, February 2012.
15. Strasser B. (2004) A Scientific Entreprenuer. *American Scientist*, March 2004.
16. Horowitz NH, Berg P, Singer M, *et al.* (2004) A Centennial: George W. Beadle, 1903–1989. *Genetics* **166**: 1–10.
17. ECF interview with Paul Berg, February 2012.
18. Berg P. (2006) Origins of the Human Genome Project: Why Sequence the Human Genome When 96% of it is Junk? *Am J Hum Genet* **79**: 603–605.

CHAPTER TWENTY NINE

Public Policy Issues — And Other Interests

Freedom to conduct scientific enquiry is inherent in the right to free speech

S uzanne Pfeffer, the current chair of the Stanford Department of Biochemistry, proudly claims some responsibility for getting Berg involved in public policy work — or at least maintaining that involvement. "As a member of the American Society of Cell Biology's council he was elected chairman of the Public Policy committee. He accepted the invitation even though he did not consider himself a cell biologist," Pfeffer related. "I believe that he was looking to be involved with public policy issues in which he thought he could make a difference. These so-called retirement years were just as successful as his science years. He found a new stage and used it effectively."[1]

"He was very active with Congress that absolutely loved him," Pfeffer continued. It would appear that Congress's love for Berg was (sometimes) reciprocated by him. In October 1996 he wrote to Congressman John Porter of the House of Representatives:

Dear Congressman Porter,

I write to commend you and thank you for your initiative and decisive role in getting the 1997 fiscal year through the Congressional Appropriations minefield. You are a wizard, one to whom every biomedical scientist owes a debt of gratitude.[2] ··

357

As an aside, it bears mention that Berg's epistles to congressmen were by no means always congratulatory. His scathing invective never too deeply buried, in December 1998 he strayed from science and public policy by dropping the following note concerning hearings on the Monica Lewinsky affair in a congressman's lap:

The Honorable Tom Campbell
United States House of Representatives
Washington, D.C. 20515

Dear Representative Campbell,

I am not one of your district's constituents but as a Stanford colleague I must express my dismay and disappointment at your votes on the articles of impeachment: two up and two down is most unconvincing, even waffling.

I heard you defend the article charging perjurious testimony in the Paula Jones deposition on the grounds that it deprived her of a fair hearing in her sexual harassment charge. It was clear to all who were open-minded that her case was activated three years after it was alleged to have occurred, financed by those who wanted the President "dismembered" and in the end was rejected by the court and ended by having Jones paid off!! To cap it off, the N.Y. Times reported your vote as being against the second article, the very one that you said was even more egregious than the alleged perjury before the grand jury!! What hypocrisy![3]

I hope that the resentment you incurred in your district is not forgotten at the next election time.

Returning to Berg's positive interactions with the US Congress, Suzanne Pfeffer commented: "He was one of the few people who could explain basic science in a way that they understood. He really understands how to communicate."[4]

This was indeed (and still is) an important chapter in Berg's life, during which he has come to realize how much difference he can make behind the scenes, with regard to funding for biomedical research and other controversial issues in science — such as religion and science. He also became involved with the Coalition for Life Science, a partnership of scientific organizations that contributes financially to the Cell Biology Society. Pfeffer made special mention of Berg's leadership roles at Stanford — especially in the biochemistry department. "Very engaged," she emphatically stated.[5]

****

In the 1980s, the DNA community of scientists embarked on a new venture: the Human Genome Project (i.e. obtaining the entire sequence of the human genome), which involved Berg. Its initiation was not without considerable controversy — this time, whether or not to advance such a formidable undertaking. This argument was not about risks. Rather, at the height of the debate the concerns centered on whether the information to be gained would be valuable and worth the price. After all, it was obvious to all that determining the precise order of the entire three billion base pairs would be very costly — but the intellectual value of having the sequence was indeterminate.[6]

Like the recombinant DNA controversy, the history of the Human Genome Project has been the subject of numerous books and articles and will not be addressed here in any detail. Readers familiar with this milestone in human endeavor, one often compared to landing a man on the moon, are presumably aware of the enormous benefits realized and still promised from the project, not to mention the (correctly) predicted progressive reduction in the cost of sequencing DNA, to the point that sequencing the genomes of entire human populations is now a seriously considered undertaking.

The first that Berg heard of the project was when he read an editorial in *SCIENCE* written by Renato Dulbecco in 1986 strongly advocating that the scientific community undertake this important work — principally because it would reveal all the "cancer genes." Berg was initially skeptical. "But at some point I thought, why not? Even if looking for cancer genes might not tell us as much as Dulbecco had thought, I realized from my experience with bacteriophages and the mammalian tumor viruses how valuable it had been to know their genomic sequences."[7]

Several months prior to a forthcoming 1986 Cold Spring Harbor Laboratory Symposium on the Molecular Biology of *Homo Sapiens*, Berg contacted Jim Watson, director of the Cold Spring Harbor Laboratory on Long Island, NY and the primary organizer of the symposium. Berg suggested to Watson that it might be useful to identify a small group at the meeting to discuss this issue. When he arrived at the Cold Spring Harbor Laboratory, Berg discovered that Watson had already arranged such a meeting and suggested that Walter (Wally) Gilbert and Berg lead the discussion. But instead of this being a rump session of a small group, Watson had scheduled this event for the opening of the meeting for all to hear — and the auditorium was packed. Berg continued the story:

> As the attendees assembled it was clear that the project was on the minds of many, and almost everyone who attended the symposium showed up for the session at the newly dedicated Grace Auditorium. Wally Gilbert and I were assigned the task of guiding the discussion. Needless to say, what followed was highly contentious; the reactions ranged from outrage to moderate enthusiasm — the former outnumbering the latter by about five to one.
>
> Gilbert began the discussion by outlining his favored approach: fragment the entire genome's DNA into a collection of overlapping fragments, clone the individual fragments, sequence the cloned segments with the then existing sequencing technology, and assemble their original order with appropriate computer software. In his most self-assured manner, Gilbert estimated that such a project could be completed in 10–20 years at a net cost of ~$1 per base, or ~$3 billion. Even before he finished, one could hear the rumblings of discontent and the audience's gathering outrage. It was not just his matter-of-fact manner and self-assurance about his projections that got the discussion off on the wrong foot, for there was also the rumor (which may well have been planted by Gilbert) that a company he was contemplating starting would undertake the project on its own, copyright the sequence, and market its content to interested parties. One could sense the fury of many in the audience, and there was a rush to speak out in protest. Among the more vociferous comments, three points stood out:
>
> 1. The cost of doing this project would diminish federal funding for individual investigator-initiated science and thereby would shift the culture of basic biological research from "Little Science" to "Big Science." Some feared that biology would experience the same

consequence that physics did when massive projects like the Stanford Linear Accelerator Center were undertaken in that field.

2. Many thought that Gilbert's approach was boring and thus would not attract well-experienced people, which; most likely, would make the product suspect. Moreover, the benefits of the sequence project might not materialize until the very late stages.

3. A surprisingly vocal group argued that because <5% of the DNA sequence was informational (i.e. represented by genes encoding proteins and RNAs) there was no point in sequencing what was unaffectionately labeled "junk" — defined as all the stuff between genes and within introns. Why, many asked, should we spend a lot of money and effort to sequence what was clearly irrelevant?[8]

Berg was "blown away" by the hostility of the response from the audience, much of which he blamed as an expression of antagonism to Gilbert's perceived haughtiness — at least in Berg's own mind. But what irked him most was a pervasively arrogant sentiment that other than the nucleotides that encoded amino acids for proteins, learning a genome's entire sequence was probably irrelevant, a sentiment based on the pervasive opinion that knowing the sequence of the 95% of the genome that did not encode amino acids was dispensable. There was also major concern that the cost of such a project, anticipated as being in the billions of dollars, would siphon funds for the support of traditional basic research. Berg tried mightily to encourage the group to focus on the logistics of mounting a Human Genome Project — assuming that the work had no impact on existing grant funding. But the assembled scientists were not forthcoming and the meeting quickly degenerated into bedlam.

As argumentative as the discussions at Cold Spring Harbor were, they prompted the US National Academy of Sciences to constitute a committee chaired by academy president Bruce Alberts to explore the issue further. In due course the committee crafted a proposal to undertake this challenge — and, notably, to include existing model organisms, such as yeast, flies, mice and other model organisms. The committee also agreed that sequencing the human genome right off the bat was not a sensible way to proceed. The appropriate technology for such large-scale DNA sequencing should be in place before such an effort was launched.

Watson and David Baltimore visited NIH director James (Jim) Wyngaarden and convinced him that supporting the project was a worthwhile effort. Watson was appointed Director of the Human Genome Project. But his agenda then was not the science — the biology. It was on technology. "One of the (few) suggestions I offered at the Cold Spring Harbor meeting was to clone and sequence cDNAs instead of genomic DNA," Berg related. "Such an approach would have the merits of both speed and a considerably reduced initial financial outlay and would reveal the nucleotide sequence of DNA that encoded proteins, arguably the most pressing information wanted," Berg stated. "But Watson's adamant focus was to first develop appropriate technology to map the genome; initially constructing a genetic map based on single nucleotide polymorphic markers (SNPs)* and then a physical map of DNA fragments without regard to the order of nucleotides they contained. Watson strictly forbade DNA sequencing financed by NIH research grants."[9]

Enter Craig Venter, an ambitious molecular biologist then at the NIH, who brazenly declared that *he* was going to independently adopt the cDNA approach, and began doing so in a private company he floated — and patenting bits and pieces of human DNA as he progressed. Bernadine Healy was then the NIH director and was supportive of the idea of cloning and sequencing cDNAs. "Her view was that sequencing the genome would put the US at the forefront of genomics," Berg stated. "Jim [Watson] was absolutely livid at this development and voiced disparaging comments about Healy in public. She in turn leveled accusations about Watson's investment portfolio that incensed him to the point that he resigned the directorship — or was fired. I don't remember which."[10]

---

* Single Nucleotide Polymorphisms (SNPs) represent a natural genetic variability at high density in the human genome. A synonymous expression is "biallelic marker," corresponding to the two alleles that may differ in a given nucleotide position in a diploid cell. A SNP represents an alternate nucleotide in a given and defined genetic location at a frequency exceeding 1% in a given population. This definition does not include other types of genetic variability like insertions and deletions, and variability in copy number of repeated sequences. SNPs are considered to be the major genetic source to phenotypic variability that differentiate individuals within a given species. [http://ghr.nlm.nih.gov/handbook/genomicresearch/snp]

While all this drama was unfolding, Berg was asked to serve as chairman of the Scientific Advisory Board for the project, a requirement established by Congress. "At this juncture I believed that Jim had done a fine job lobbying Congress for financial support for the project," Berg stated, "so I agreed to serve in the capacity that was requested, although I was seriously concerned that the advisory board had no director to report to. At a working dinner the evening before the first meeting of the board I pushed the cDNA idea once again. No one was interested; perhaps because now the word on the street was that Sydney Brenner was urging the powers that be in the UK that cDNA sequencing was the way *they* should proceed."[11]

In the final analysis Berg attended no more than two or three board meetings before a decision was made that the composition of the board could not include more than a single individual for each institute, a directive that placed David Botstein (then a faculty member at Stanford) on the committee. "Since David seemed much more in tune with what was going on than I, it seemed logical that I should go off the committee," Berg related with evident lack of concern. "So I was on the Advisory Board for no more than about a year."[12]

That was the extent of Berg's involvement in the Human Genome Project. He didn't bring about any change except perhaps loosening Watson's strong hand as director. Scientists involved in the project were then complaining about Wally Gilbert's idea of fragmenting the genome into multiple pieces and systematically cloning them — so-called shotgun cloning — and then solving a jigsaw puzzle of millions of fragments. Many, including Berg, did not believe that gifted postdoctoral fellows, and most certainly graduate students, would be pleased to cut and sequence bits and pieces of DNA devoid of any biological context. "That was another reason why I had argued strongly for the cDNA approach. If a postdoc cloned and sequenced a cDNA he or she would at least have something tangible to explore immediately."[13]

****

The most recent controversy emanating from biomedical science, one that is as bitter if not more so than that tied to the use of recombinant

DNA because it embraces religious views about the sanctity of life, is popularly referred to as the "stem cell controversy." All multicellular organisms possess so-called stem cells in organs and tissues in which cells normally divide. As the name implies, stem cells are progenitor cells that can multiply, thereby giving rise to mature cells in a tissue or organ. Stem cells are functionally *undifferentiated*, i.e. they are not endowed with attributes that support specific functions, such as do muscle cells, liver cells or nerve cells, examples of so-called *differentiated* cells. Such specialized (differentiated) cells are not able to spawn generations of new specialized cells by simple cell division. Hence, when for example muscle, liver or nerve cells die as a result of disease or injury, or even old age, they can only be replaced by stem cells, which have an essentially unlimited capacity for renewal. In adult organisms, stem cells are able to differentiate into the same type of adult cells that they replace. Thus, stem cells in the liver can replace lost liver cells, but no other type of lost cells. In contrast, stem cells from embryos are *totipotential*, i.e. they can differentiate into all sorts of adult cells. Consequently, if one wishes to replace lost liver cells one must either provide adult stem cells from the liver, or preferably (in terms of relative technological ease) cells derived from human embryos that are about to be discarded.

The prospect of using stem cells to repair or replace damaged or diseased organs has, not surprisingly, generated enormous attention in the general public. Some scientists speculate that such cells may even be able to functionally replace an entire diseased organ, a notion that spawned the emergence of a new medical field called *regenerative medicine*.

Aside from their totipotentiality, embryonic stem cells are readily available from embryos, many of which can be harvested from *in vitro* fertilization (IVF) procedures — instead of simply discarding them, a waste of potentially valuable material from which to generate embryonic stem cells. Furthermore, whereas much is known about how to promote the differentiation of embryonic stem cells, in contrast, adult stem cells not only have limited differentiative capacity, they are additionally much more difficult to differentiate in the laboratory. But many hold the view that even though very young embryos are unable to survive for long outside the womb, they are nonetheless living entities and should be protected by the same moral, ethical and religious principles that protect older living organisms. Hence the controversy!

In 1996, the US Congress approved prohibition of the use of federal funds for research involving the creation or destruction of human embryos. Most gratifying to the proponents of the use of embryonic stem cells, in 1999 the US Department of Health and Human Services (HHS) ruled that while federal monies may not be spent for research on human embryos, such research involving stem cells *previously* derived from human embryos was legal. Accordingly, two years later, on August 9, 2001, President George W. Bush announced that embryonic stem cell research may proceed — but only with stem cells collected prior to that date.

At first blush this proclamation seemed encouraging. However of the 66 embryonic cell lines established and maintained in government laboratories, essentially all proved to be unusable for one reason or another. So the issue of deriving new stem cell lines from embryos surfaced once again — with renewed vigor and with renewed polarization of opinions. To advise him on the matter of stem cell research, President Bush appointed Leon Kass (who the reader will recall was an early proponent of Berg's research on the development of recombinant DNA; see Chapter 16) to chair a new Ethics Committee.

When Bill Clinton held the office of President he was not in principle opposed to stem cell research. However, he recognized that the issue would be highly provocative. So he instructed Dr. Shirley Tilghman, a well-known molecular biologist and president of Princeton University to chair a committee to draft guidelines. The committee produced a white paper that Tilghman asked Berg to review. "I was a strong supporter of stem cell research, mainly because I firmly believed in the enormous scientific and medical promise of a new area of biology," Berg stated. "I gave a lot of talks and lectures on the subject — shocking numerous people in the process." At a talk at Yale University in September 2004, Berg pulled no punches. "How could any administration with any conscience at all prohibit the possibility of saving lives just because the technology offends them?" he blistered.[14]

In due course Berg was asked to testify before a Senate Committee, where he confronted the chairman of the hearing (who was a physician) by asking him how he would feel about having to inform a patient who had a life-threatening disease that might benefit from stem cell therapy — given that he had voted to prohibit stem cell research.[15]

In 2006, Berg was invited to present the Hogan and Hartson Jurimetrics Lecture in Honor of Lee Loevinger at the Arizona State University Sandra O'Conner College of Law. The lecture was subsequently published in *Jurimetrics*, a journal of law, science and technology, under the title "Brilliant Science, Dark Politics, Uncertain Law."[16] In addition to remarks on the recombinant DNA controversy, Berg devoted much of his lecture to the ongoing stem cell controversy.

Relief for the position of those in favor of embryonic stem cells research appeared on the horizon when in 2001, Robert Klein II, a Palo Alto real estate developer whose son was afflicted with type II diabetes and whose mother was suffering from Alzheimer's disease, approached Berg. Klein was outraged by President Bush's prohibition of federal funding for human embryonic stem cell research and, as he and others saw it, forestalling research that might help cure his son and others afflicted with type II diabetes. "Klein was both wealthy and bright," Berg related. "And he was well informed about the ins and outs of writing legislation authorizing the raising of bonds."[17]

Klein rolled up his sleeves and spearheaded an energetic campaign to place a proposition on the California election ballot in 2004. Berg devoted considerable time and energy campaigning in favor of the proposition. "I suspect I got sucked into this effort because it was well known locally that I had experience with this sort of public policy work," he explained.[18] Proposition 71, the *California Stem Cell Research and Cures Initiative,* was passed by a handsome majority and became an amendment to the California constitution. The proposition authorized the sale of $3 billion of general obligation bonds and created the California Institute for Regenerative Medicine to disperse $300 million a year for 10 years. Funds were to be disbursed based on review of written proposals from qualified academic institutions in California. Klein also established an independent governing committee, the Independent Citizen's Oversight Committee

Robert Klein II. (THE 2005 *TIME* 100.)

(ICOC) to oversee the process. In 2010, with a grant of $47 million from the California Institute for Regenerative Medicine, the Institute for Stem Cell Biology and Regenerative Medicine was added to Stanford University Medical Center. At the time of this writing, Stanford stem cell authority Irving Weismann directs this enterprise.

In light of the ethical debate that accompanied the introduction of embryonic stem cell research, several private biomedical research entities chose not to provide funding. These passive opponents included the Susan Komen Foundation — and more distressingly, the American Cancer Society (ACS). The latter organization attracted the attention of an irate Paul Berg, who wrote to the CEO of the Society in August 1999:

> .... the failure of the ACS to join with other scientific and disease groups in providing full support for federally funded investigators to work in this area is shocking. The public reports justifying this decision implied that it was taken to mollify a small number of the Society's financial supporters who oppose this line of research on ideological grounds. It seems to me that the ACS with its long and distinguished record of support of cutting-edge biomedical research would have rejected this pressure without hesitation. The ACS's failure to take such a principled stand on this issue represents a betrayal to patients and their families, many of whom have supported the Society's efforts to pursue every lead and discovery in the quest for a cure to the cancer scourge. ....[19]

368 A Biography of Paul Berg

In early 2013, the US Supreme court provided a huge boost to stem cell research by refusing to hear an appeal that would have abolished federal funding for embryonic stem cell research.

In the course of the heated debate on embryonic stem cell research, Berg composed a brief essay entitled "Galileo Redux" in which he argued that freedom to carry out research was being threatened, not because it represented a clear and present danger to human or environmental health, but because it was an affront to the personal ethics or ideology of some. He voiced the opinion that a line of otherwise legitimate investigation aimed at the acquisition of new knowledge and the development of novel therapies for very serious diseases was at risk of being criminalized for offending a segment of the public's personal taste or values! He also warned that the quality of scientific investigations may no longer be the sole or principal determinant in pursuing a particular line of research; rather, theological and ideological considerations parading as fundamental ethical values may increasingly take over. "The chilling effect on our scientific enterprise is palpable," he wrote.[20]

The "proper relationship between society and science" that surfaced in the wake of the stem cell polemic troubles Berg deeply — as it does millions of others. In a 2005 piece published in the *San Jose Mercury News* bearing the title "Fearing to go where science leads — Right to inquire, freedom of expression go hand in hand," Berg raised the question as to whether or not federal interference with basic research is a violation of the right to free speech granted in the US Constitution's Bill of Rights. "I believe the case can be made that the freedom to conduct scientific enquiry is inherent in the right to free speech ...... And to abridge that implicit right bodes ill for science and our democracy as well." he wrote.[21]

****

As pointed out earlier, Berg is drawn to contentious issues, scientific or otherwise, like a moth to a flame. He is especially adept at expressing his opinions in writing, frequently in a letter or editorial format associated with a prominent scientific journal. But, as already noted, he does not hesitate to write directly to the sources of some of his concerns.

In early 1993, Berg was prompted to challenge film director and producer Steven Spielberg about aspects of his film *Jurassic Park*. Cognizant of the imperative to dissociate himself from the mass of anticipated trivial comments to Spielberg from members of the general public, Berg identified himself as "one of the founders of the science of genetic engineering ·············· and currently Chairman of the National Advisory Committee on the Human Genome Project" and informed Spielberg of his desire to "comment on the polemical aspects of [Michael] Crichton's work":

> Your efforts to dramatize the unimaginable on the screen are widely praised and admired. However, on scientific grounds, the premise that prehistoric animals could be reconstructed by patching together bits of DNA recovered from insects embedded in amber, though clever, is patently absurd; indeed, assembling genomes of even far simpler organisms from highly fragmented DNA may be impossible, especially if only variable parts are recoverable.
>
> What worries me about Crichton's book, and the possibility that it has been perpetrated in the film, is the quite blatant assertion that current researchers in molecular genetics are unmanaged, and that through commercial enticements they are quite capable of malevolent and destructive practices. Indeed, his brilliant young geneticist resembles a stand-in for Dr. Strangelove, and is intended, I believe, to promote the kind of abhorrence that was engendered by the prospect of an insane use of nuclear weapons.[22]

Aside from the fact that it is not obvious why Berg chose to communicate with Spielberg rather than Michael Crichton, a trained physician who was then still living, it's frankly surprising to encounter this sort of "huffiness" from someone with Berg's intelligence, sensitivity and insight — not to mention his obvious sense of humor. Perhaps 18 years after the recombinant DNA controversy was put to rest he remained genuinely concerned that public perception of "mad scientists" may encourage further recombinant DNA and embryonic stem cell sagas. More likely Berg was either suffering from intense boredom, or his prolific writing hand was especially itching for action that February day! Regardless, after waiting patiently (and ultimately in vain) for a response from Spielberg, Berg managed to obtain one from a Spielberg assistant — who promptly told him to "fuck off!"[23]

As if membership on Schultz's Hoover Institution Energy Task Force, reading the scientific literature, and providing advice and council on an *ad hoc* basis didn't keep the 86-year-old Berg sufficiently busy, he also sat on a committee of Sr. Associate Deans at the Stanford Medical School that met once a week under the chairmanship of Dean Pizzo. Not surprisingly, in 2008, Berg was a recipient of the Dean's Medal, presented by Pizzo at a dinner celebrating the school's centennial. "Without question, ------- Professor Paul Berg has been essential in shaping the medical school and medical center as we know it today and in preparing it for the challenges and successes it will face in the decades to follow," Pizzo said in announcing the award.[24]

****

Berg and his wife Millie have for many years devoted considerable time, energy — and money — to collecting art, a topic that lights up his eyes. "During a time when we lived in 'hovels' we always had things on the walls," he proclaimed. "I would say my appreciation for *good* art developed when we went to Paris on the way to Copenhagen in 1952. I remember going to the *Jeu de Paume* and viewing work by the French impressionists for the first time. While in Copenhagen we visited numerous museums and art galleries, but we had no money then to indulge in any serious collecting."[25]

When the Bergs moved to California from St, Louis in 1959, a friend educated them to appreciate high quality original lithographs, etchings and drawings. Slowly and cautiously the Bergs began buying works by local artists from the Smith Andersen Gallery in Palo Alto. One of the local artists was Nathan Oliveira, an art professor at Stanford. At the time Oliveira was more focused on making monotypes. Berg was intrigued by the technique and by the images Oliveira was creating. So "we bought some of his work." This remarkable and world-renowned artist died in his home in Stanford on November 13, 2010 at the age of 81.

In due course, the Bergs met a dealer in San Francisco named John Berggruen. John's father, Heinz Berggruen, had immigrated to the

United States prior to WWII. He married a prominent San Francisco woman and joined the US Army at the start of the war in Europe. At war's end Berggruen established an art gallery and exhibited many upcoming post-impressionist artists, among who were Picasso and Braque. Indeed, at one time he was perhaps one of the most eminent art dealers in Europe. His son John grew up in San Francisco and eventually also became an art dealer. Over time the John Berggruen Gallery has become one of the leading art galleries on the west coast and one of the Bergs' favorite places to visit when in San Francisco. "John taught us a lot about collecting and frequently helped us manage costs when our tastes exceeded our budget," Berg said.[26]

"One of our more interesting acquisitions transpired when Millie fell in love with an artist named William Bailey, who at the time was Dean of the Yale University art school, but also maintained a studio and dealer in Rome, Italy. Any thoughts about acquiring a Bailey work were stymied by their scarcity and price, but we learned that there was to be a Bailey exhibition in Rome at the time we planned to be there during our return from the Stockholm Nobel celebrations."[27] Providentially, Berg had an Italian postdoctoral fellow in his laboratory and asked him to contact the gallery's dealer to request the individual to save several of the works in the show for his arrival. "Surprisingly, he did so and when we arrived in Rome we chose a piece — now one of our favorites." Over the years the Bergs have amassed a fine collection of works by contemporary American artists. Their single misfortune is the theft of a watercolor by Wayne Thiebaud — that was stolen off the wall of the gallery after they had paid for it! "Over the years we have acquired a pretty good art collection. Once we had covered all the walls of our house we began collecting sculpture."

Paul and Millie once had the celebrated artist Nathan Oliveira (mentioned above) to a dinner party at their home. The other guests included a filmmaker from Hollywood who was on a sabbatical at Stanford and who was quite taken by the art displayed on the walls. After dinner, Oliveira, the filmmaker and Berg walked through the house examining and commenting on the art collection. "The filmmaker suggested that it would be interesting to make a film about two people from very disparate fields who share a single common interest — and to capture the conversation between them. A graduate student in filmmaking was duly commissioned to make the film, which ultimately turned out to be his PhD thesis,"[28] Berg related. The result was a short film in which Berg and Oliveira roamed the hills behind Stanford talking about making art and doing science, their similarities and differences. The film showed scenes from one of Berg's regular laboratory staff meetings, juxtaposed with Oliveira working through the process of creating a painting and a monotype. What a novel interlude in an aging scientist's life!

The late Henry S. Kaplan, who as mentioned several times was the former chair of the Department Of Radiology at Stanford and a pioneer of employing radiation for treating cancer, was also a passionate art collector. When in 1979 Berg wrote to Kaplan congratulating him on receiving a prize from the General Motors Cancer Research Foundation, he suggested that the famous radiotherapist spend his prize money on art. Kaplan responded:

> That [thought] had already occurred to me. In view of your own beautiful art collection, I would be glad to appoint you my art consultant, in case you see anything that you think would be of interest.[29]

Notes

1. ECF interview with Suzanne Pfeffer, September 2012.
2. Letter to Congressman John Edward Porter, October 2, 1996. Reproduced with permission from the Department of Special Collections, Stanford University Libraries.
3. Letter from Paul Berg to the Honorable Tom Campbell, December 1988.
4. ECF interview with Suzanne Pfeffer, September 2012.
5. *Ibid.*
6. Berg P. (2006) Origins of the Human Genome Project: Why Sequence the Human Genome When 96% of it is Junk? *Am J Hum Genet* **79**: 603–605.
7. ECF interview with Paul Berg, September 2012.
8. *Ibid.*

9. *Ibid.*
10. *Ibid.*
11. *Ibid.*
12. *Ibid.*
13. *Ibid.*
14. *Ibid.*
15. *Ibid.*
16. Berg P. (2006) Brilliant Science, Dark Politics, Uncertain Law. *Jurimetrics* **46**: 379–389.
17. ECF interview with Paul Berg, September 2012.
18. *Ibid.*
19. Letter to John Seffrin, August 24, 1999. Reproduced with permission from the Department of Special Collections, Stanford University Libraries.
20. Berg P, unpublished essay.
21. San Jose Mercury News, January 20, 2005.
22. Letter to Steven Spielberg, February 16, 1993. Reproduced with permission from the Department of Special Collections, Stanford University Libraries.
23. ECF interview with Paul Berg, September 2012.
24. http://med.stanford.edu/mcr/2008/dean-medal-0402.html
25. ECF interview with Paul Berg, February 2012.
26. *Ibid.*
27. *Ibid.*
28. *Ibid.*
29. Letter from Henry Kaplan to PB, May 9, 1979. Reproduced with permission for the Department of Special Collections, Stanford University Libraries.

CHAPTER THIRTY

Personal Challenges

The Berg family is a small but tightly knit trio. Paul and his son John often eat out together, as do Paul and Millie. John Berg and his wife, Kris, shared Paul and Millie's 50th wedding anniversary in Italy. Away from the laboratory and office, Berg is an interesting, convivial and entertaining companion — or host, with little hesitancy in striking up idle conversation, even with perfect strangers who happen to be seated close by in public situations. In many ways, Berg is a prototypical American product of the mid-1940s and the 1950s, the proverbial straight arrow, informally dressed, honest and modest to a fault. But there is also a "jock" in Berg. He has always loved (and still does) sport, both as a participant and as a spectator. The turbulent decades of the 1960s, '70s and '80s have left their mark too, exposing his strong liberal streak and his passionate concern for societal issues outside his professional expertise. A man with eclectic interests, he is, besides his keen interest in art, a bookworm.

But recent years have challenged both Paul and John (not to mention Millie) with the stress and anxiety that inevitably accompany serious illnesses. John, an extraordinarily charming and engaging personality, bears no obvious physical resemblance to either of his parents, but is blessed with an impressive appearance in his own right — and a lively and charming disposition. "John was a good kid," Paul Berg related. "As a two year old he talked so incessantly that Millie and I could only converse when he was asleep!"[1]

As he grew older it was obvious that John was bright and lively, and Millie and Paul had little trouble with him — until he reached his teenage years in the late 1960s and early 1970s. John's many passions (which now include tearing around on his state-of-the-art motorcycles) encompass music, which has captivated him since his boyhood. As a middle school student he was unusually enthralled by and well-versed in the history of jazz and rock and roll. By his own initiative he was offered a freewheeling midnight to 6am shift as a licensed disk jockey at the Stanford University radio station KZSU. "Millie has never forgiven me for allowing him to take this job, because he was at an age of trying to emulate the older guys — and that led him into a troublesome period," related Berg.[2]

An extraordinarily eclectic individual, John has (and still does) nurtured many interests from an early age, and Paul and Millie were delighted to discover that their son has been extraordinarily talented in essentially everything he tried his hand at.[3] Highly independent and certainly conscious of his father's considerable fame, John was predictably determined to carve out a niche of his own. When he was at high school in Palo Alto, his biology teacher (who was of course well aware of Paul's stunning accomplishments) once patronizingly informed him that she "expects great things from you" — at which point John promptly dropped his biology class!

Surprisingly, however, John followed in his father's footsteps by enrolling at Penn State University, where he mainly focused his formidable energy on acting. "I'm sure that he spent much of his time in drama studies and little in other courses," Berg commented.[4] In due course John's passion for the stage gained him acceptance into the *Neighborhood Playhouse*, a famous New York-based acting school that among many others boasts as graduates the likes of Gregory Peck, Steve McQueen, Robert Duvall, Diane Keeton, Joanne Woodward, Tony Randall and Jeff Goldblum — to mention just a few.

Prior to venturing off to New York, John was contacted by the director of a Broadway production that was scheduled to be in residence at Penn State for a semester, during which they would be performing *The Shadow Box*, a play written by Michael Cristofer that made its Broadway debut in 1977. The play won both the Tony Award and the Pulitzer Prize that year.

The director had apparently seen John perform during the previous year and was keen to audition him for one of the parts set aside for a student. But John decided to forgo enrolling at the *Playhouse* and returned to Penn State, where he won a lead role in *The Shadow Box*. With that experience behind him and having completed his second year at Penn State, John returned to the *Neighborhood Playhouse* for a year and was poised to launch a career as an actor — when he fell in love with a young Stanford student (Chris Bondurant) whom he met before leaving for New York. The couple subsequently married, but this union lasted less than a year, and they went their separate ways. A year or so later John married Kris Schmidt, an elementary school teacher in Palo Alto. They have a daughter, Stephanie, a dancer who returned to the Bay Area after graduating from Mt. Holyoke.

With both emotional and material support from his parents, John has thrived in his own eclectic existence, one with an astonishing range of interests and accomplishments as a swim instructor, musician, stage actor and photographer. He currently teaches and manages the *Samyama Yoga Center* in Palo Alto.

In early 2000 at age 42, John discovered a small, painless lump in his neck that was diagnosed as B-cell lymphoma. The emotional odyssey he and his parents have experienced has been a predictable one and bears no description here. Having endured the punishing chemotherapy and radiation treatments that so often follow the diagnosis of cancer, John is totally symptom free at the time of this writing. But like all cancer survivors, especially those still in the prime of life, he is always on the alert. He talks of this experience as "the best thing that ever happened to me in terms of completely shaking up my priorities — I mean REAL priorities."[5]

As mentioned earlier in this narrative, Paul was for many years an ardent (and by all reports) an able tennis player. However, the frequent vigorous exercise, coupled with the stresses and strains of aging, necessitated surgeries for spinal stenosis, repair of a damaged mitral valve and a coronary bypass, and a hip replacement. He recovered sufficiently from each of these episodes to return to work and daily activities. But the discovery of a cancerous lesion in his left leg in 2006 (initially diagnosed as malignant melanoma) proved to be a considerably more challenging experience. Numerous surgeries necessitated multiple tissue grafts. Additionally, radiation to the tumor and a consequential fracture of the tibia required two successive stressful surgeries to promote reunion of the fractured bone, and left him with a need for a cane to help navigate his surroundings — with surprisingly little discomfort or handicap! Thankfully, Berg has suffered no further complications nor does he seem concerned at the possibility of the cancer recurring.

While in the midst of this saga, Berg and colleagues at Stanford were able to grow cells from his tumor in tissue culture, characterize their surface proteins and perform gene expression and genomic profiling. Their data revealed that the tumor was not a malignant melanoma, as was first diagnosed; the findings were more consistent with a diagnosis of sarcoma. There is little question that many (perhaps most) others fortunate enough to have survived a protracted succession of medical challenges in their mid-80s would have by now adopted a life style that kept one mainly at home. Paul Berg, however, uncomplainingly drives his car or his scooter to his office — and back home every day.

In ending Paul Berg's life story to date it would be remiss to not mention the role his wife Millie has played in making it all possible. The long and odd hours that Paul's scientific research consumed as well as the long absences for lecture tours, meetings and writing retreats placed a strain on her otherwise caring personality. John refers to his mother as the "Rock of Gibraltar" for her unflappability when life's challenges have struck. Expert nursing care was her gift to all. Except for one surgery, soon after John was born, and a relatively recent fractured vertebra, Millie has held off the usual infirmities that strike women of her age.

On November 3, 2011, Stanford University president John Hennessy organized a small and exclusive reception in Paul Berg's honor. Paul concluded his obligatory remarks of thanks with a tribute of his own — to Stanford University, where he has spent the last 54 years of his life. "Now, I am more of an observer than a participant in the university's activities. Nevertheless, one cannot miss the throbbing energy and drive towards

Paul and Millie Berg — the sunset years. (Courtesy of Paul Berg.)

still greater challenges and accomplishments. There is the vision, the talent and commitment, and not to be minimized, the financial support from so many to make those dreams a reality," Berg related with obvious sincerity.[6]

Notes

1. ECF interview with Paul Berg, June 2012.
2. *Ibid.*
3. *Ibid.*
4. *Ibid.*
5. ECF interview with John Berg, April 2011.
6. ECF interview with Paul Berg, June 2012.

References

Appel TA. (2000) *Shaping Biology. The National Science Foundation and American Biological Research 1945–1975*. The Johns Hopkins University Press.

Beckman J. (2002) *Making Genes, Making Waves. A Social Activist in Science*. Harvard University Press.

Berg P, Singer M. (1992) *Dealing With Genes: The Language of Heredity*. University Science Books.

Berg P, Singer M. (2003) *George Beadle, An Uncommon Farmer: The Emergence of Genetics in the 20th Century*. Cold Spring Harbor Laboratory Press.

Ciba Foundation Symposium 31 (New Series. In tribute to Fritz Lipmann on His 75th Birthday). (1975) *Energy Transformations in Biological Systems*. Elsevier Publishing Company.

Denniston KJ, Enquist LW. eds. (1981) *Recombinant DNA Benchmark Papers in Microbiology/15*. Dowden, Hutchinson & Ross.

Denson C. (2002) *Coney Island: Lost and Found*. Ten Speed Press.

Feldman B. (2000) *The Nobel Prize: A History of Genius, Controversy, and Prestige*. Arcade Publishing.

Florkin M, Stotz EH. eds. (1972) *Comprehensive Biochemistry, Vol. 30. A History of Biochemistry*. Elsevier Publishing Company.

Fredrickson DS. (2001) *The Recombinant DNA Controversy: A Memoir: Science, Politics and the Public Interest*. ASM Press.

Fruton JS. (1972) *Molecules and Life: Historical Essays on the Interplay of Chemistry and Biology*. John Wiley & Sons, Inc.

Friedberg EC. (2010) *Sydney Brenner — A Biography*. Cold Spring Harbor Laboratory Press.

Fruton JS. (1977) *Selected Bibliography of Biographical Data for the History of Biochemistry Since 1800 (2nd edition).* American Philosophical Society Library Publication No. 7.

Goodfield J. (1977) *Playing God: Genetic Engineering and the Manipulation of Life.* Harper & Row.

Grobstein C. (1979) *A Double Image of the Double Helix: The Recombinant DNA Debate.* W. H. Freeman and Company.

Hanna KE. ed. (1991) *Biomedical Politics.* National Academy Press.

Hargittai I. (2002) *Candid Science II: Conversations With Famous Biomedical Scientists.* Imperial College Press.

Hellman A, Oxman MN, Pollack R. eds. (1973) *Biohazards in Biological Research: Proceedings of a Conference Held at the Asilomar Conference Center,* Cold Spring Harbor Laboratory Press.

Hughes SS. (2011) *Genentech: The Beginnings of Biotech.* University of Chicago Press.

Jackson DA, Stich SP. eds. (1979) *The Recombinant DNA Debate.* Prentice Hall Inc.

Jacobs CD. (2010) *Henry Kaplan and the Story of Hodgkins Disease.* Stanford University Press.

Judson HF. (1979) *The Eighth Day of Creation: The Makers of the Revolution in Biology.* Simon and Schuster.

Kohler RE. (1982) *From Biomedical Politics to Biochemistry: The Making of a Biomedical Discipline.* Cambridge University Press.

Kalckar HM. (1969) *Biological Phosphorylations: Development of Concepts.* Prentice-Hall Inc.

Kiernan V. (2006) *Embargoed Science.* University of Illinois Press.

Kornberg A. (1989) *For the Love of Enzymes: The Oddyssey of a Biochemist.* Harvard University Press.

Kornberg A. (1995) *The Golden Helix — Inside Biotech Ventures.* University Science Books.

Krimsky S. (1982) *Genetic Alchemy: The Social History of the Recombinant DNA Controversy.* The Massachusetts Institute of Technology.

Lear J. (1978) *Recombinant DNA: The Untold Story.* Crown Publishers, Inc.

Leicester HM. (1974) *Development of Biochemical Concepts: From Ancient to Modern.* Harvard University Press.

Lipmann F. (1971) *Wanderings of a Biochemist.* John Wiley & Sons, Inc.

Merton RK. (1973) *The Sociology of Science: Theoretical and Empirical Investigations.* The University of Chicago Press.

Morgan J, Whelan WJ. eds. (1979) *Recombinant DNA and Genetic Experimentation: Proceedings of a Conference on Recombinant DNA Jointly organized by the Committee on Genetic Experimentation (COGENE) and the Royal Society of London, Held at Wye College, Kent, UK, 1–4 April.* Pergamon Press.

Morgan RM. (1976) *The Genetics Revolution: History, Fears, and Future of a Life-Altering Science.* Greenwood Press.

National Academy of Sciences. (1977) *Research With Recombinant DNA: An Academy Forum, March 7–9.* National Academy of Sciences.

Needham J. ed. (1971) *The Chemistry of Life: Eight Lectures on the History of Biochemistry.* Cambridge University Press.

Park A. (2011) *The Stem Cell Hope: How Stem Cell Medicine Can Change Our Lives.* Hudson Street Press.

Proceedings of the Conference on the History of Biochemistry and Molecular Biology May 21–23 (1970). American Academy of Arts and Sciences.

Rogers M. (1977) *Biohazard: The Struggle to Control Recombinant DNA Experiments, the Most Promising (and Most Threatening) Scientific Research Ever Undertaken.* Alfred A. Knopf.

Rosenfelt L. (1999) *Four Centuries of Clinical Chemistry.* Taylor & Francis.

Scott WA, Werner R. eds. (1977) *Molecular Cloning of Recombinant DNA, Miami Winter Symposia,* Vol. 13. Academic Press, Inc.

Singer M, Berg P. (1997) *Exploring Genetic Mechanisms.* University Science Books.

Singer M, Berg P. (1991) *Genes & Genomes: A Changing Perspective.* University Science Books.

Vettel EJ. (2006) *Biotech. The Counterculture Origins of an Industry.* University of Pennsylvania Press.

Wade N. (1977) *The Ultimate Experiment: Man-Made Evolution.* Walker and Company.

Watson JD, Tooze J. (1981) *The DNA Story: A Documentary History of Gene Cloning.* W. H. Freeman and Company

Watson JD, Berry A. (2003) *DNA: The Secret of Life.* Alfred A. Knopf.

Wright S. (1994) *Molecular Politics: Developing American and British Regulatory Policy for Genetic Engineering, 1972–1982.* University of Chicago Press.

Yi D. (2008) Cancer, Viruses, and Mass Migration: Paul Berg's Venture into Eukaryotic Biology and the Advent of Recombinant Research and Technology. *J Hist Biol* **41**: 589–636.

Zilinskas RA, Zimmerman BK. (1986) *The Gene-Splicing Wars — Reflections on the Recombinant DNA Controversy.* American Association for the Advancement of Science. Macmillan.

Index